Penetrant Testing

Penetrant Testing

A practical guide

David Lovejoy

Kluwer Academic Publishers
DORDRECHT / BOSTON / LONDON

Published by Kluwer Academic Publishers,
P.O. Box 17, 3300 AA Dordrecht, The Netherlands

Sold and distributed in North, Central and South America
by Kluwer Academic Publishers,
101 Philip Drive, Norwell, MA 02061, U.S.A.

In all other countries, sold and distributed
by Kluwer Academic Publishers,
P.O. Box 322, 3300 AH Dordrecht, The Netherlands

First edition 1991

ISBN 0 412 38700 X (PB)

British Library Cataloguing in Publication Data

Lovejoy, D.J. (David J.)
Penetrant testing.
1. Materials. Testing
I. Title
620.1127

ISBN 0-412-38700X

Library of Congress Cataloging-in-Publication Data
Available

Contents

Preface

It is about ten years since I completed the manuscript for this book and first I should like to thank all of the readers of it who have told me how useful they have found it and indicated that it has been largely successful in its objectives. During the intervening years the actual technology of penetrant testing has changed very little however other aspects of its application have brought changes. Perhaps the simplest to comment upon is that the greater availability of and reduced costs of information technology has begun to open the way to a new generation of automatic installations. Other sources of changes are from either environmental concerns or health and safety aspects or a combination of these. Developments in both areas have had a profound influence on the use and applications of industrial chemicals in all areas however the influence on penetrant testing is fairly specific.

The effects of the loss of availability of 1, 1, 1, tricholoroethane to penetrant testing has led to the use of other chlorinated solvents for pre-cleaning and the use of flammable solvents in the actual application of the penetrant processes. These changes are discussed fully in sections in Chapters 3, 5, 6, 7, 8, and 10. The problem of drying surfaces properly after pre-cleaning with water based chemicals before the application of penetrant and a possible management of the problem are discussed in Chapter 5 where a new section "5.1.10 Drying surfaces" has been added. The current concerns about all volatile organic compounds is mentioned however it is nog possible to predict any developments in the next ten years. In Chapter 10 the concerns about exposure of the skin to UV. A which has become more widely known during the past ten years are discussed with suggested actions for anyone who is concerned about this.

If volatile organic solvents as a class do become restricted during the coming years this will present a challenge to the suppliers of penetrant materials and there will be inevitable changes in the way that this most useful method of NDT is applied.

This book is intended to give a concise guide to the penetrant inspection method of non-destructive testing for all operators, inspectors, supervisors and engineers whose work involves these techniques. The design of the book is to give a useful manual for both training and day-to-day use. The emphasis is practical and is aimed at indicating how the various process can be applied supported by enough background to explain why. I hope that this approach indicates the very great scope that penetrant testing offers for finding surface-breaking defects when a suitable process is chosen for a problem and is properly applied.

Penetrant testing is sometimes considered as a poor relation among non-destructive testing techniques. This is largely because it produces results which, while they are readily seen and recognized, are extremely difficult to record. The fact that it can be a little messy also destracts from its reputation, as does the fact that the basic process is so simple in outline. The greatest contributor to the relatively low status sometimes given to penetrant testing is that it has often been carried out poorly and the results have not been as reliable as they should and could have been.

Much of the problem arose from the fact that, until 10 years ago, formal training is the proper application of penetrant testing was not commonly required in industry and many operators may or may not have been formally trained. This practice had a number of shortcommings: first, and most obvious, if the original operator had a faulty technique this was passed on to the learner; second, the operator's experience was limited to one or very few techniques. Training in the penetrant method of non-destructive testing has grown enormously and continues to expand. Excellent schemes such as PCN (United Kingdom) ASNT (United States) DGZFP (Germany), COFREND (France), APIND (Italy) and others have helped to formalize training and expand practitioners' knowledge of specific non-destructive testing methods and their contribution to quality control programmes.

The aim of this book is to offer a basic grounding in practical penetrant testing. The first chapter gives a brief historical survey of the history of technology and the introduction of non-destructive testing. The origins of defects is a topic which has inspired some excellent books in its own right, and it is dealt with here to give the background to the need for testing. Chapters 3–7 deal with the penetrant process itself, its development, principles of the process, techniques, choice of method and the equipment which can be used. The quality control of penetrant materials and processes is dealt with in Chapter 8, and specification and documentation are covered in Chapter 9. The final two chapters deal with health and safety and the management of penetrant waste respectively. These last chapters involve some technical terminology which may be completely new to some readers. However, this cannot be avoided.

I sincerely hope that users of the penetrant method of non-destructive testing find this book to be a useful manual through the various stages of training in this technology.

I should like to thank the many people who, over the past years, have helped me to understand better the problems associated with penetrant testing and in their discussion helped to indicate the potential in it. I should like to thank specifically Messrs R.E. Birley, N.H. Hyam, M. Perry, B.C. Graham, A.M. Fijalkowsky, R. Selner, J.T. Schmidt, J. Vaerman, J.-L. Meifrenn, M. Gillard, E. Dickhaut and S. Mastellone for many stimulating discussions concerning penetrant materials and their applications.

1
Introduction

Without technology there would be no need for non-destructive testing (NDT) and consequently no need for penetrant inspection methods. It is useful to take a brief look at the development of technology to the present day when millions of people take for granted such impressive results of that development as modern aircraft, space travel and the fact that light and electrical power is available in people's homes at the turn of a switch. This introductory chapter traces the outline of technological development, use of materials and energy sources from early times. This short summary leads to the need for and introduction of NDT methods, thus emphasizing the importance of the work of anyone involved in this field. Without the various methods of NDT many of the benefits of modern life would be unavailable, and every operator, inspector, technician, scientist and engineer should be aware of the importance of their work and the need for it to be carried out to the highest standard at all times.

1.1 A BRIEF HISTORY OF TECHNOLOGY

As soon as man used tools our species had embarked on the road of technology. The earliest human relics are 1.75 million years old; they were discovered by Mary Leakey in the Olduvai Gorge in East Africa, and a significant fact is that the rock from which these earliest known tools were made does not occur locally in the area of settlement but had to be brought from a considerable distance. The materials available to our earliest ancestors include stone in various forms, plant products such as wood, reeds, leaves and grasses, and animal products such as bone, antlers, horns, skin, hair and sinew.

The use of iron is known to have become established in a number of areas of the world by 3000 years ago. The Bible refers to the fact that skills in the use of iron were in the hands of the Hittites and Philistines, which caused the Children of Israel problems from time to time, the Greek poet Hesiod regretted the coming of the Iron Age as long ago as the seventh century BC, and by the time the year 239 BC was reached Chinese metal workers were discussing the composition of bronzes and its effect on brittleness and elasticity.

The museums of the world show many examples of man's artefacts from earliest times to modern times. Even without contemporary written documents the advanced skills of the technologists of all ages are self-evident. Where technical documents exist, as is the case with early Chinese, Greek, Persian and Roman work, these show that the designers and makers of these artefacts thought about their problems as much as modern man does today.

The decline of the Roman Empire saw the transfer of developing technology to the Muslim world which adopted Greek and Roman knowledge and built on it. Several of our names for scientific and mathematical studies are Arabic in origin, the best known being algebra and chemistry. During the 1000 years following the year AD 500 most, if not all, technological advance reaching Europe came from the east. Windmills made their first appearance in Europe in the eleventh century whereas they had been in use in Persia for 400 years by then. It is very significant that the first recorded rolling mill in Europe was in Segovia in Spain where it was used for the manufacture of strips of gold and silver for coins. The significance of the fact that it was in Spain is that the Moors, a Muslim people of mixed Arab and Berber race, had occupied that country for many years and had left their technology there.

After the year 1500 technology in Europe advanced at an ever increasing pace in common with the many other activities which form the basis of the Renaissance. Within 250 years of this date the first steam engines were actually working in Europe and by 1830 the revolution in transport took physical shape with Robert and George Stephenson's railway engines transporting paying passengers. Steam engines were not simply used for rail transport; they had been in use industrially for more than 100 years before the first passenger railway was opened. Boats, machine tools and agricultural equipment all benefited from the invention and development of steam engines.

In 1860 Lenoir built and demonstrated the first internal combustion engine which used coal gas as fuel; 2 years later another Frenchman proposed a four-stroke version of this engine but neither publicized the idea nor tested it experimentally.

Nikolaus Otto, the German inventor, studied Lenoir's invention and by 1876 had produced an engine giving 3 h.p. at 180 rev/min using a four-stroke cycle. Karl Benz used this principle in the cars he was selling by 1890.

The prospect of flying has intrigued man for at least 3000 years. Around 1000 BC the Chinese were using kites to carry military observers, just as European military forces did in the nineteenth century. However, kites are fixed to the ground and the amount of freedom for the occupant is restricted.

The first manned aerial flight was made by Pilatre de Rozier and the Marquis d'Arlandes who flew 5 miles across Paris on 21 November 1783 in a balloon designed by the Montgolfier brothers. Modern flight was born many years later in 1905 at Kitty Hawk in North Carolina when the Wright brothers succeeded in their efforts at true flight with *Flyer III*.

In 1939 a centrifugal flow jet engine designed by Dr Hans von Ohain successfully powered a Heinkel HE178 aircraft. This German design and that of Sir Frank Whittle in the United Kingdom opened the era of the jet engine.

Technology had come a long way from stone tool manufacture in eastern Africa by the slow steady accumulation of knowledge.

1.2 DEVELOPMENT OF MATERIALS

Stone, plant products and animal products were the only materials available to early man. It is interesting to note that stone and plant products still feature as important materials today for the manufacture of tools and for structural uses.

The ideas of technologists are often more advanced than the materials available. There are many examples of this, one of the most famous of which must be Leonardo da Vinci's designs for aircraft. In specific terms the steam turbine is an excellent example of this fact. In the second century AD the Greek Hero worked out the principle of this and in 1629 Giovanni Branca revived the idea. Neither had materials which would tolerate the speeds of rotation; in 1884 Charles Parsons did have suitable materials and he was able to produce a steam turbine which rotated at 18 000 rev/min, and so a power source for driving ocean liners and generating electricity was born.

The first materials other than those from stone, plant and animal sources were metals such as copper, gold and silver, which are found in the free state, and their alloys (bronzes) with tin and zinc, which are easily extracted from their ores. This was followed by the extraction of iron by smelting. It was not possible for the commonest metal in the earth's crust, aluminium, to be used until quite recently as electric current is needed to extract it from its ores. The percentage by weight of the commonest eight metals in the earth's crust is given in Table 1.1. Old records show a remarkable volume of empirical knowledge of extraction metallurgy and alloy technology with a vast range of organic materials being recommended for smelting iron and for steel manufacture.

Table 1.1 Abundance of metallic elements in the earth's crust

Element	*Percentage by weight*	*Element*	*Percentage by weight*
Aluminium	8	Potassium	2.6
Iron	5	Magnesium	2
Calcium	3.6	Titanium	0.4
Sodium	2.8	Manganese	0.1

All other metallic elements are present in abundances of less than 0.1 wt%; for example, copper is present at 0.0045 wt%.

Stainless steel was first discovered in 1911 simultaneously in Germany and the United Kingdom. Until relatively recently the stainless steels have been regarded as special and expensive materials. While their cost remains higher than that of mild steel, their use is now widespread and they are certainly no longer considered 'special'. Titanium, once thought of as a very special material, is used extensively in the aircraft industry; it is still costly but by no means uncommon.

Plant tissues are polymers and many of them, such as wood and string, are still used. The modern polymer industry began in 1850 with the discovery of celluloid which was originally called Parkesine after its inventor Alexander Parkes. Celluloid was followed by other early synthetic polymers including Bakelite. Today a vast range of polymers has been developed including adhesives, packaging film and structural materials. The special properties of composites are now widely exploited in the structure of aircraft. To a significant extent polymers can be designed for their intended use.

The oldest known technological material of all, stone, has of course been used for various purposes throughout the history of man. In modern times the development of ceramic materials, which are stone based, has been enormous and the value of these materials, while usually unseen, is important in many industries.

The choice of materials in today's world is very wide indeed. In 20 years' time it will be wider still and some materials which are important now will have been replaced by new ones. The special properties of single-crystal metals, composites, polymers and ceramics will allow specific design and choice of materials to ensure ever greater efficiency in all industries.

1.3 ENERGY SOURCES

Table 1.2 summarizes the development of technology from the earliest times to the present day.

Muscle power, either human or animal, wind and moving water have been available sources of mechanical energy throughout the entire history of man. The only controllable source of energy which uses gravity and does not require obvious use of one of the other forms of energy to complete a cycle is a waterfall. Evidence of the potential of the basic sources of energy is with us still in the shape of such structures as the Pyramids in Egypt and Central America, Stonehenge in the United Kingdom and other stone circles and alignments. An average modern man can develop a power of 100 W and for a short time this can be doubled. When the oval type of collar is used a horse or an ox can develop 1600 W. Many machines using muscle power have been developed including treadmills, capstans and winches. Hand tools, other than the power-driven type, are still operated entirely by muscle power.

Water wheels are known to have been in use in the Far East over 2500 years ago, and the use of tides stretches at least from the third century AD,

Table 1.2 An outline of the rise of technology

Date	*Technology/materials*	*Energy*	*Transport*
1.75 million years ago	Stone tools from quartzite, lava and other rocks in Olduvai Gorge, East Africa	Muscle Wind Water	
3000–2000 BC	Early Bronze Age Copper alloys with tin and arsenic in use		Sailing vessels on water
2000–1600 BC	Middle Bronze Age		Wheeled vehicles for land use
1600–1200 BC	Late Bronze Age		
1000 BC	Transition to Iron Age		Improved transport on land and sea
700 BC	Hesiod regrets coming of Iron Age		
500 BC	Evidence of steel making in China		
239 BC	Discussion of bronze metallurgy for swords in China		
AD 500	Decline of Roman Empire – start of Dark Ages in Europe	Muscle Wind Water	Boats and wheeled carts
AD 700	Windmills known to be working in Persia (Iran)		
AD 1100	Windmills reach Europe		
AD 500 onward	Rapid development of technology in the Muslim world from Alhambra to Samarkand		
AD 1500	Rolling mill working in the mint at Segovia, Spain, probably of Moorish origin		
1500–1850	Early work and principles of electricity established		
1712	Newcomen's steam engine	Steam	
1794	Ballbearing invented	Muscle Wind Water Steam	Marine craft River craft Animal-drawn vehicles
1804	Early railways		
1830	Passenger railway		Railways using only steam
1833	Water turbine		
1860	Lenoir's internal combustion engine	Hydro-carbon	Primitive car
1864	Driving chains		
1884/6	True motor car		
1893	Diesel engine		
1904	Caterpillar track		
1905	First aeroplane		
1907	Helicopter		
1911	Stainless steel		
1938–9	Jet engine		
1942	Nuclear power		

when a tide mill is known to have been in use in Dover, to the modern use of tides for power stations as in France. The first true water turbine design was a prize-winning entry for a competition held by the French technological society in 1826. Variations of Fourneyron's winning design are in use today.

The use of wind to propel boats is known to have a history of at least 5500 years. The use of wind for windmills is more recent. These were developed in Persia in the seventh or eighth century AD and were used for irrigation. The prototype is thought to be the prayer wheels which have been common in Asia since the fifth century AD.

Early methods of power generation are shown in Fig. 1.1.

The possibility of using steam as a source of power was considered by many of the great minds of history. Archimedes and Hero in ancient Greece and later Leonardo da Vinci gave a lot of thought to the problem. Otto von Guericke in Magdeburg, Germany, Denis Papin in France and Thomas Savery in England all developed the ideas further but with very limited success. In 1712 Thomas Newcomen built the first practical steam engine with much help from John Calley, Robert Hooke and the Swedish scientist Morton Triewald. Forty years later James Watt and Thomas Boulton improved the Newcomen design and by 1800 more than 500 of their engines were in use, in Boulton's words to King George II 'uplifting civilisation by relieving man of undignified drudgery'. Improvements to the Boulton and Watt design, leading to higher working steam pressures, were made by many engineers including Trevithick in the United Kingdom and Evans in the United States.

Later development of steam engines and the subsequent development of jet engines owes its theoretical basis to the work of the Frenchman Sadi Carnot. Carnot published his principles of heat engines in 1824. He suffered from the fact that his explanation was not understood by most of his contemporaries. Ten years later another Frenchman Benoit Clapeyron, expanded Carnot's theory and tried to explain these principles, but he too suffered from a lack of understanding on the part of other engineers and scientists. In the 1840s two Germans, Robert Mayer and Robert Clausius, and the Englishman James Joule developed and explained the work of Carnot and Clapeyron so that it could be readily used by all professional engineers. By the end of the nineteenth century steam turbine design was well established.

Lenoir showed that the energy from exploding coal gas could be used to produce mechanical energy. Otto and Benz extended this to petroleum, and in 1893 Rudolf Diesel introduced the compression ignition engine which bears his name. These developments gave the world power sources for motor cars, commercial road vehicles, boats of all sizes, aircraft and a wide range of industrial vehicles.

Gas turbines are, in principle, very simple designs. The problems of actually making them were associated with available materials and a proper

Fig 1.1 Early methods of power generation: wind and water were two of the first sources of power used to replace muscle power, either human or animal. Current concern for the environment has encouraged a re-evaluation of the potential of these energy sources.

understanding of Carnot's work. By the end of the nineteenth century the Norwegian engineer Elling had produced a gas turbine which was the ancestor of the engines that von Ohain and Whittle succeeded in developing for use with aircraft and through them the range of modern power plant for civil and military jet aircraft.

Static electricity was known to the ancient Greeks. Recent excavations in Iran have uncovered artefacts which are probably electric cells. Modern understanding of electrical energy and application of it are due to Galvani,

Volta, Davy, Oersted, Ampère, Ohm, Faraday, Clerk Maxwell, Siemens and Edison. Imagine a day without electricity and our debt to these men becomes clear.

When Rutherford first split the atom he opened the way for the use of nuclear power. The ideas of Otto Hahn, Lisa Meitner and others led to the development of the first nuclear reactor by Enrico Fermi's team in 1942. The fact that this enormous power was first released as a result of military development is a reflection of mankind rather than technology. Even as the terrifyingly destructive military devices were being developed and used many people dreamed of and planned for peaceful use of this newly released power source. While not yet universally popular nuclear power sources already play an important role in supporting modern civilization.

As sections 1.4 and 1.5 show, the extension of technology, the use of more materials and the exploitation of more energy sources increases the need for all forms of quality control including NDT.

1.4 THE DEMANDS OF TECHNOLOGY ON MATERIALS

On 10 November 1840 at the bottom of the Lickey incline on the Birmingham and Gloucester Railway the boiler of the locomotive *Eclipse* exploded killing the driver and fireman. In May 1864 three very destructive boiler explosions occurred on three different railways. On 5 May 1864 the boiler of a goods engine at Colne on the Midland Railway exploded killing the driver and injuring the fireman, and an elderly lady lying in bed in her cottage a quarter of a mile away was struck on her leg by a piece of plate from the explosion which fell through the roof. The boiler of locomotive number 138 of the Great Northern exploded on the morning of 9 May 1864 demolishing most of Bishop's Road Station on the Metropolitan line. In the afternoon of the same day the boiler of engine number 108 exploded at Leominster on the Shrewsbury and Hereford line. Fortunately in the last two cases there were no casualties but there was very substantial demolition of property. A less spectacular but often more disastrous source of problems was fracture of wheels, tyres or axles and cracking of the permanent way. Wheel fractures occurred at Southall in 1847 on Brunel's broad gauge and casualties were light. Six years later there was a similar breakage on the standard gauge Lancashire and Yorkshire Railway with much more disastrous results with the driver and five passengers being killed. These early disasters were followed by many more. At Penistone on New Year's Day 1885 the axle of a wagon on a goods train broke just as a passenger train was passing in the opposite direction. This unfortunate coincidence cost four passengers their lives. In the nineteenth century Railway Companies became depressingly familiar with collapsing bridges. In 1852 the steamship *Reindeer* was leaving dock at Malden on the Hudson River when the boiler flue exploded. The 26 deaths ensuing were due to scalding. Eighteen months later the boiler flue burst again on the same boat killing more than 40 people by its direct and indirect effects.

Static boiler explosions were more spectacular. On 2 March 1854 at the Fales and Gray Car Works a newly installed boiler exploded killing 21 people, injuring 50 more and demolishing a brick wall of dimensions 12 metres × 24 metres. In the early 1920s a boiler exploded in Canton, Ohio. The fireman's body was blown 150 metres, at which point it passed through a house and demolished a fence on the other side. The boiler itself took off in another direction shedding rivets during its flight.

Such spectacular and public evidence of the failure of materials and structures could hardly go unnoticed, and if people wished to continue to work with steam at high pressure or to travel at high speed such as 50 or 60 miles/hour companies would have to take measures to stop killing employees, passengers or passers-by and stop destroying other people's property. Also, the loss of their own property and ensuing loss of produciton capacity was a good stimulus to seeking remedies. After all, explosions of steam-driven road coaches on British roads in the 1850s led to their being banned and created a political hostility to road transport which still persists in some quarters even today.

1.5 INTRODUCTION OF NON-DESTRUCTIVE TESTING METHODS

From the beginning of technology it is reasonable to assume that some form of visual inspection was carried out. The settlers in the Olduvai gorge 1.75 million years ago were intelligent enough to survive, and intelligent enough not only to make tools but also to use various types of rock and to travel some distance to obtain these materials. Such people were unlikely to lack the intelligence to look at what they had made – to analyse breakages and to seek to avoid them in future. The very survival of these people depended on their tools, and visual inspection must have played its part even if no specifications have been left as to procedures. Visual inspection has certainly played an important part in the recorded history of technology and still does. Aided by 5×, 7× and 20× magnification visual inspection is still made of critical parts such as turbine blades for jet engines. The process is slow and laborious. After 2 hours inspectors need rest as their concentration starts to decline. Despite the disadvantages of this method it still retains an important place in modern NDT.

The wholesale destruction caused by exploding boilers and trains derailed by broken wheels, tyres, axles or track called for a marked increase in the use of visual inspection both in the construction of equipment and during overhaul. Although this was a good step forward, the limitation of the fact that visual inspection can only find surface defects operated then as now. It was soon found that tapping methods –the precursor of ultrasonic testing – could detect many fractured wheels and tyres on the railway. It could also be used for the estimation of wall thickness, thus finding eggshell-thin boiler plate before it exploded. The oil and whiting method, which is a crude form of penetrant testing, was first applied to railway axles and later to boiler plate.

In the late 1920s fusion welds replaced rivets as the means of joining boiler plate. The development of magnetic particle testing, electrical induction testing and radiography developed rapidly with consequent improvement of safety and performance. By 1945 recognizably modern ultrasonic testing of welds had become a reality.

The success of early steam power with its associated technology, notably on the railways, highlighted the need for NDT. By the time airflight was practicable a fair body of NDT knowledge existed, and the need to avoid structural failure or engine failure while in the air is very obvious. The rise of the aerospace industry in the 1930s and during the Second World War was matched by assimilation of known NDT techniques and development of new techniques to match the needs of the new technology. Similarly, and equally fortunately, the development of the nuclear energy programme started at a time when NDT had become established. The effects of mishandling this source of energy were so clear that, as in the case of the aircraft industry, awareness of the need to use NDT existed from the start. At present there are six widely used NDT methods:

1. visual inspection
2. magnetic particle inspection
3. radiography
4. penetrant testing
5. eddy current testing
6. ultrasonic testing

To this list we can add more recent techniques such as acoustic emission and holography. It is interesting to note that no NDT technique which is firmly based scientifically has ever been dropped, but only modified and added to. The spectrum of NDT methods continues to grow in line with the demands of technology and its advances.

2
Origins of defects

From the very outset of the use of metal for tools, implements, weapons, containers, transport or any other type of artefact man has had to find ways of overcoming two basic problems. One is corrosion and the second, which concerns us here, is failure.

The failure of metals must have been very apparent from the very outset of the Bronze and Iron Ages. The advantages of metal artefacts in the early stages of their development must have obscured the disadvantages, since alternative materials such as stone, horn and, in many applications, wood suffered even more problems. The problems of failure must have become apparent and even embarrassing as man became more ambitious in his development of metals. Spear heads which cracked when striking a hunted animal's hide instead of piercing it must have been a cause of friction between hunter and early iron manufacturers. Swords which sheared off in battle must have been a severe embarrassment to their owners and it is doubtful if they were in a position to discuss the matter with their suppliers later. Early attempts to reduce the incidence of such failures were based on the principle that more is better and weakness was to be cured simply by increasing the mass of parts or devices which failed frequently in use. That this method is basically unsound yet gained respectability until the last 100 years is a reflection on the relatively low demands made on metal components until the Industrial Revolution was far advanced and not confirmation of this approach.

As demands on metal structures and artefacts have risen so has the study of how metals behave in service, and now the use of masses threefold, fourfold, fivefold or even sixfold excess to requirements is recognized as an understandable but mistaken attempt to solve the problem of safety and service. The study and application of stress analysis has allowed careful design of metal structures and parts, but it is the intelligent application of NDT which has allowed these studies to be translated into manufacturing reality.

The history of the 'more is better' approach to the problem shows many examples of the situation being made worse by the addition of greater mass to a part by the inclusion of 'strengthening members', i.e. structures leading to local stress concentrations leading in turn to early and frequently catastrophic failure.

Early studies of failure were not restricted simply to increasing mass but also paid much attention to actual materials. It is both easy and fashionable to scorn the early alchemists, but their achievements in what is now called metallurgy were considerable. Early mysterious methods for hardening and alloying metals were certainly developed by trial and error, but once established were carefully followed. The manufacturer may not have known why certain steps were necessary to produce his wares, but certainly from the days of the Egyptian Pharoahs to the middle of the last century a strong developing knowledge of how to produce hardened and tempered steels and a wide variety of other useful alloys for many uses existed. To a very great extent this depth of empirical knowledge of handling metals compensated for the almost total lack of proper study of design.

As the pace of the modern machine-oriented age gathered momentum so the requirements on materials increased. Fortunately studies of design stress and metallurgy attracted more attention and effort. The development of metallography gave the metallurgist, for the first time, the ability really to look into the structure behind that hitherto mysterious wall, the metallic surface. At last it was possible to study the metallic microstructure and learn why metals behave in the way they do and to see the effect of hot working, cold working, tempering and all the other methods of modifying metallic characteristics which had been developed by thousands of forgotten metal workers from Asia, Africa and Europe over thousands of years. The studies of the internal structure of steels and other metals quickly led to an understanding of how and why metals fail. Differentiation between the effects of static and dynamic loading was recognized. Discontinuities which were found to be quite harmless under static loading were found to lead to catastrophic failure under dynamic conditions. Simply making the metal stronger by heat treatment or alloying failed to improve the incidence of service failure and often made it worse. This type of observation stimulated the modern study of stress and strain which gives us an understanding of the role of design in failure-free service.

2.1 MECHANISMS OF FAILURE

Metals and other materials fail in a number of ways. Two major mechanisms are brittle fracture and ductile fracture.

1. Brittle fracture is said to occur when parts break with very little or no discernible plastic flow.
2. Ductile fracture is said to occur when the break is preceded by a very significant amount of plastic flow.

Both brittle and ductile fracture cut across grain boundaries. Both types follow slip lines within the grains of the metal; however, these slip lines are on different axes of the crystal in the cases of the two types of fracture.

Conditions leading to failure include the following:

1. simple overstressing following overloading;
2. impact induced by a sudden stress rise;
3. fatigue induced by stress variations below the elastic limit;
4. corrosion;
5. creep

2.1.1 Overstressing

All materials have a limited strength and if they are loaded beyond this they pull apart and fail. Overloads of this type occurring in service have several causes. Unusual conditions may cause overstressing. For example, a violent storm can cause failure in one part of a structure which in turn places an unusual extra load on the rest of the structure creating a stress too great for the remaining structure to carry. Similarly, sudden stoppage of moving machinery or vehicles can cause overstressing in parts of the structure. Failures from this cause are frequently of the ductile type, although some very hard materials may display brittle fracture in such circumstances and the brittle mechanism may also be seen at low temperatures.

2.1.2 Impact

In many ways impact is a further form of overstressing – the principal difference is the rate at which the stress is applied. A sudden impact can impose very high stress at a very rapid rate – failure generally occurs in the brittle mode. If the rate of application of stress is rapid enough, brittle fracture will occur in normally ductile materials.

2.1.3 Fatigue

Fatigue failure was by far the most puzzling type of failure to understand before the advent of metallography and even during the early stages of its study. It was rapidly recognized that such failure occurred when parts were subjected to reversing or widely varying stresses even though these might be well below the elastic limit of the metal. Equally puzzling was the fact that fatigue failure occurred suddenly at stress well below the strength of the metal. The appearance of the actual fracture, showing a smooth portion with concentric markings followed by a crystalline part, gave rise to the fallacy that metals become ‘tired’ in service and crystallize. Metallographic studies showed that a fatigue crack often starts at a small notch and spreads slowly to produce smooth faces as one face works against the other – when eventually the remaining sound portion fractures, it does so leaving a crystalline face.

2.1.4 Corrosion

When a metal is under tension stress and is at the same time exposed to corrosion, cracks develop and grow through the section until failure occurs. Corrosion cracking is normally due to attack on the grain boundary material and it is generally intergranular as opposed to fatigue cracking which is transgranular. Stress corrosion cracking will occur even under static loading conditions. Such stresses need not be externally applied but may be residual from previous operations such as welding or heat treatment.

2.1.5 Creep

Metals under tensile load and at high temperatures undergo slow changes which can end in failure. The metal stretches in the direction of the tensile stress. Creep data are available for many materials for design purposes. There is no creep limit, as there is in the case of fatigue, and so great care must be taken to account for this factor in design. Cracks occurring under conditions of creep can be expected to be intergranular.

2.2 STRESS RAISERS

In service, fatigue, corrosion and creep are the three major enemies of metal and other materials. Fatigue cracks, corrosion cracks and even cracks associated with creep can be seen to arise from a single point at the surface of the metal. The origin may be a scratch or even a tool mark. The tension stress across even such a small defect can be shown to be several times greater than the normal average stress at the surface.

It may not be a scratch which acts as such a stress raiser. A simple bolt hole will act in the same way. The average tension stress at a hole can be three times the average for the surface. Stiffening members added to strengthen a part may in fact act as stress raisers, thus defeating their purpose.

2.3 DEFECTS AND THEIR ORIGINS

We have already looked at some types of defect and their significance. At this point we should look at what is really meant by a defect. In common with all materials it is within the basic structure that metal defects arise. There are many words which are used in normal speech which mean the same as 'defect' or 'defective'. In terms of NDT a defect is some characteristic which renders a part or material unsuitable, unsafe or unusuable for its intended purpose. This immediately raises the problem of interpretation, for a discontinuity which constitutes a dangerous defect in one instance may be quite harmless in another. It is for the user to decide what is a defect and what is not. It is the responsibility of the NDT engineer to supply the means of detecting potential defects and to ensure that they are applied properly. This gives us a problem of definition which we solve by the use of the word 'discontinuity'. Penetrant testing is very effective in showing surface discontinuities when properly applied. Discontinuities must be open to the surface for penetrant testing to succeed. For information on the internal status of a material the NDT engineer must use other methods such as X-ray or ultrasonic testing.

There are a number of different ways in which discontinuites can be classified. They depend on location, shape, processing or other characteristics. It is useful to classify discontinuities into four broad groups:

1. produced during the initial cooling of the metal
2. primary processing
3. secondary processing
4. service

2.3.1 Discontinuities produced during cooling

Discontinuities which are present in the metal as a result of the solidification process include the following.

(a) Pipe cavities

As a molten metal is poured into moulds it cools first at the bottom and walls of the mould. Solidification proceeds upwards and inwards. Since the solid metal occupies less space than the liquid there is a progressive shrinkage. As the metal at the top of the mould solidifies last, care must be taken to ensure that there is enough left to avoid a deep cavity at the top. The normal practice of cropping the top from an ingot or billet removes this primary pipe. The cropping is carried out at a specific distance from the top without inspection. In some instances the pipe extends deeper into the ingot or billet and inspection is necessary to ensure that this secondary pipe is detected. If undetected, it can be transferred to bar stock or even components (Fig. 2.1). Apart from the defect of the secondary pipe itself, it is often associated with segregation of materials in the alloy. It is very important that inspection detects secondary pipe.

(b) Blowholes

Molten metals contain dissolved gases and as the metal cools these gases are released. If this is not controlled bubbles form which remain trapped as the

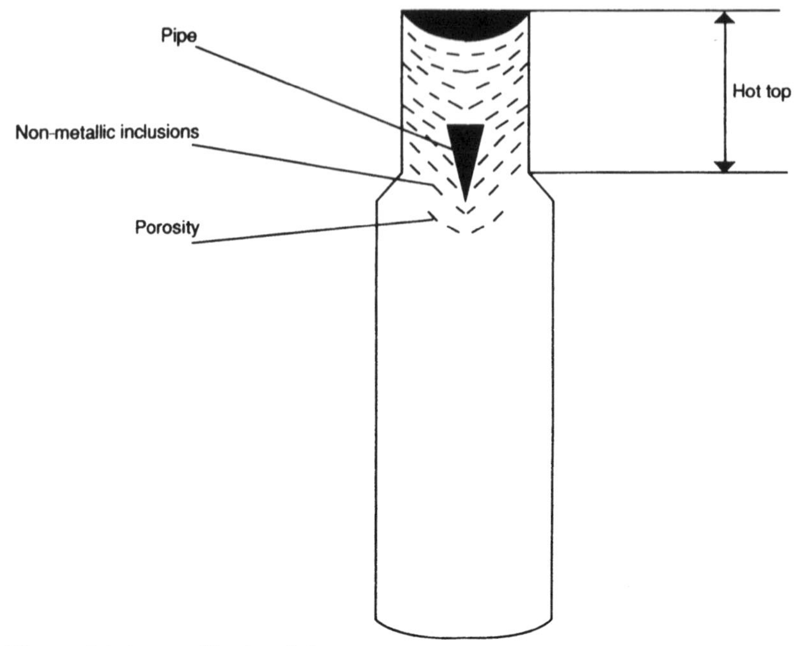

Figure 2.1 Ingot with pipe defect.

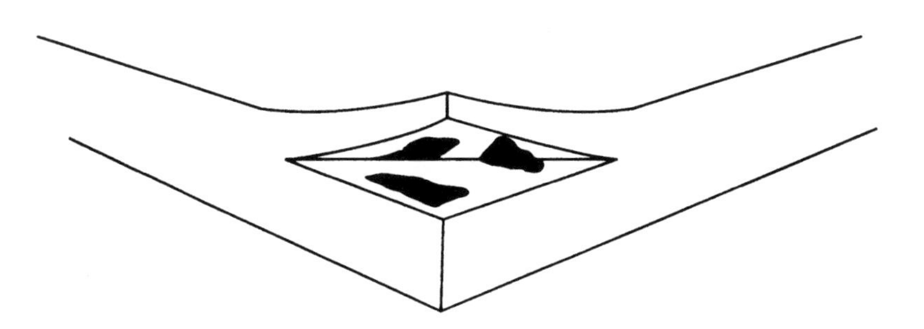

Figure 2.2 Laminations in plate material.

metal freezes. In a well-ordered procedure such bubbles will be almost exclusively at the top and will be removed when the ingot is cropped. Many of these blowholes are sufficiently clean on the inside for them to become welded together by primary processing such as rolling or forging. If the inside of the blowhole is not clean the surfaces will not weld together and the blowholes lead to the formation of seams or laminations during primary processing (Fig. 2.2).

If the cropping of the hot top does not remove the discontinuities, when the ingot is further processed into slabs the remaining discontinuities will invariably change in size and shape. As a billet is rolled into bar stock any non-metallic inclusions are squeezed out into longer and thinner discontinuities. These are called stringers (Fig. 2.3).

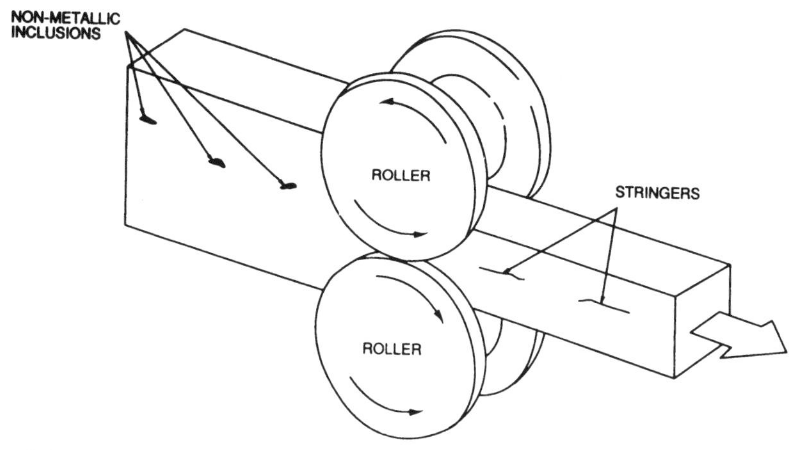

Fig. 2.3 Stringers.

(c) Segregation

Since alloys are mixtures there is a need to consider the different behaviour of the elements during cooling. Unless this is allowed for in the production process, elements can segregate. Magnetic particle inspection (MPI) will show even moderate degrees of segregation which may not be harmful to steel. Severe segregation can be a serious problem in alloys.

(d) Non-metallic inclusions

Many alloys, including all steels, contain some non-metallic material. When finely divided material is uniformly distributed throughout the alloy this is of little consequence, but such material may aggregate together and such lumps will roll out into stringers which can act as stress raisers leading to fatigue cracking in vulnerable parts.

(e) Internal fissures

As metals cool and shrink so stresses are set up. Such stresses can lead to internal fissures which can be quite large. From an NDT point of view the behaviour of such fissures is similar to that of blowholes. Provided that the surface fissures are not exposed to air and remain clean they will weld together on primary working. If the surfaces become oxidized such fissures will lead to problems later.

(f) Scabs

When liquid metal is first poured there is considerable splashing against the cool walls of the mould. Such splashes solidify rapidly and their surfaces become oxidized. As molten metal rises in the mould most of these splashes become reabsorbed into the metal, but in some cases they will remain as scabs of oxidized metal on the surface of the ingot. Provided that these do not go too deep into the ingot they will be removed on working and cause no serious problem.

(g) Ingot cracks

Just as stresses created during cooling can lead to internal fissures, so surface stresses can produce surface cracks. A major difference is that such surface cracks will certainly become exposed to air and unless removed will roll out into long seams.

2.3.2 Primary processing discontinuities

When metal ingots are worked down into billets or forging blanks some of the defects described in sections 2.3.1 will become apparent. Of course, rolling and forging can introduce their own discontinuities. Primary processing in this context includes the processes which work metals to usable forms such as bars, rod, wire and forged shapes.

(a) Seams

Seams in rolled bar stock or drawn wire are normally unacceptable. Severe seams may have their origin in the original ingot (Fig. 2.4) and can be eliminated by proper preparation of the ingots. Seams can be introduced by the rolling or drawing processes. Laps can occur if rolls are overfilled – fins form which, on subsequent passes, become rolled over onto the bar or billet, producing seams which open onto the surface of the bar at an acute angle. Seams can also be formed as a result of underfilled passes – these tend to be normal to the surface. Seams and die marks can be introduced in the drawing process as a result of defective dies. For some purposes the most minute surface discontinuities are cause for rejection; for others quite large discontinuities are not regarded as defects.

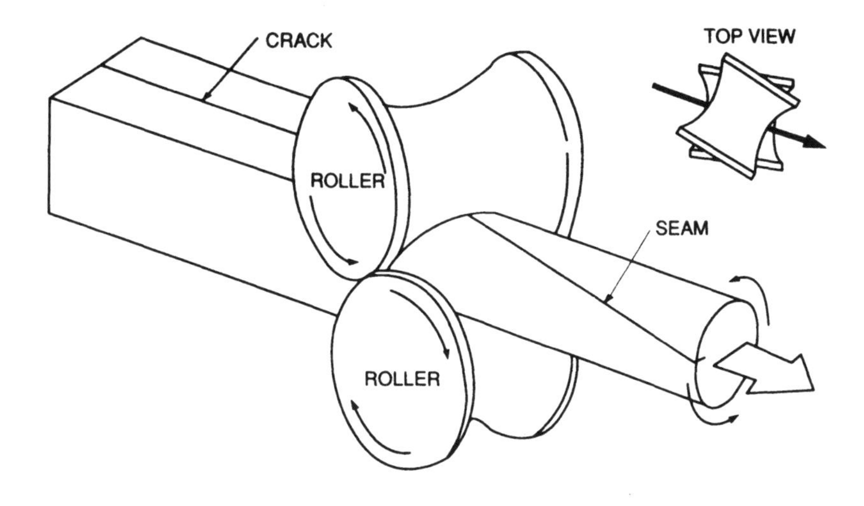

Fig. 2.4 Seam: when a billet is rolled into round bar stocks any surface irregularities may cause seams. Seams are also caused by the metal folding, due to improper rolling or by a crack in the billet. Seams can also occur when the billet is rolled from rectangular bar.

(b) Laminations

Laminations occur in rolled plate or strip when blowholes or internal fissures fail to weld tight but become enlarged and flattened into horizontal discontinuities. MPI or penetrant testing can show these on the edge of a plate or strip but ultrasonic mapping is required to define them fully.

(c) Cupping

Cupping occurs when under extrusion or drawing the interior of the metal does not flow as rapidly as the surface. Segregation at the centre of the bar contributes to this. The result is a series of severe internal defects.

(d) Cooling cracks

After rolling out, metal bars are left to cool. Uneven cooling leads to stresses which are frequently great enough to cause cracking. Such cracks are often longitudinal, though not necessarily straight, and vary in depth.

(e) Flakes

Flakes are internal ruptures which can occur as a result of too rapid cooling. They may be caused by release of dissolved gases during cooling. Flakes are internal discontinuities which can be exposed to the surface during secondary processing.

(f) Forging bursts

Forging bursts can be caused by working metals at the wrong temperature. Too rapid or too severe a reduction of section can also cause bursts or cracks (Fig. 2.5). Such bursts may be internal or they may occur at the surface.

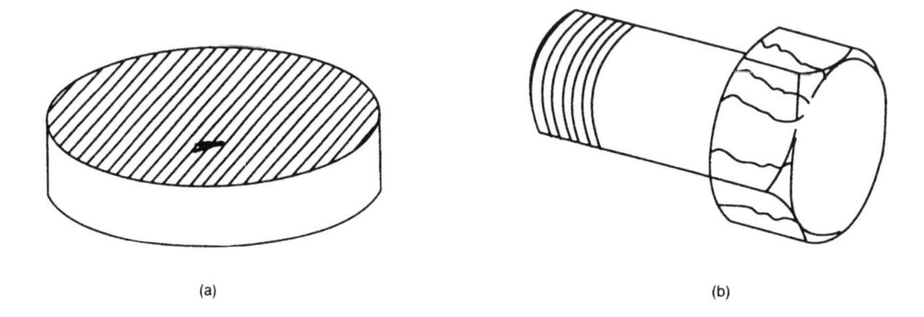

Fig. 2.5 Bursts: (a) internal (subsurface); (b) external (open to the surface). A forging burst is a rupture and is caused by forging at incorrect temperatures. Bursts can be either internal or surface discontinuities.

(g) Forging laps

As the name implies forging laps or folds are formed when the forging blank is not handled properly in the die and forms a lap which becomes squeezed tight at a later stage (Fig. 2.6). Since this is at the surface it will not weld tight.

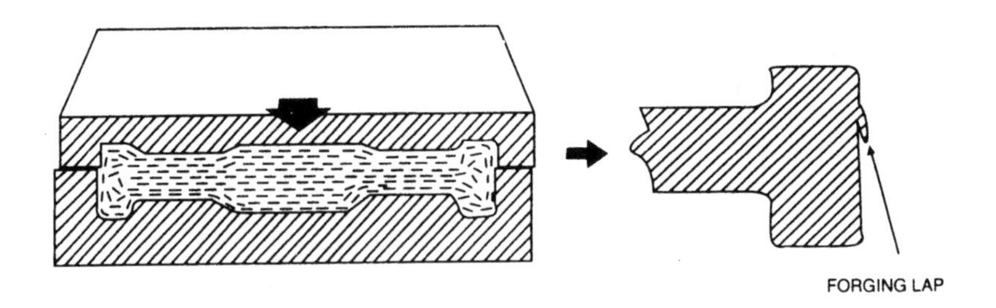

Fig. 2.6 Forging lap.

(h) Burning

Overheating of forgings to the point of incipient fusion causes a condition known as burning. Despite the name, oxidation is not the source of the problem; it is the partial liquefaction of the metal at grain boundaries. This serious defect is not normally shown by MPI or penetrant testing.

(i) Flash-line tears

Cracks or tears along the flash line of forgings are often caused by improper trimming. When shallow they are easily machined off, but deep cracks or tears are normally a serious problem.

(j) Casting discontinuities

The process of casting can lead to a wide variety of discontinuities. These include porosity, shrinkage cracks, hot tears, non-metallic inclusions (sand from a mould), cold shuts and cracks due to rough handling (Figs 2.7–2.12).

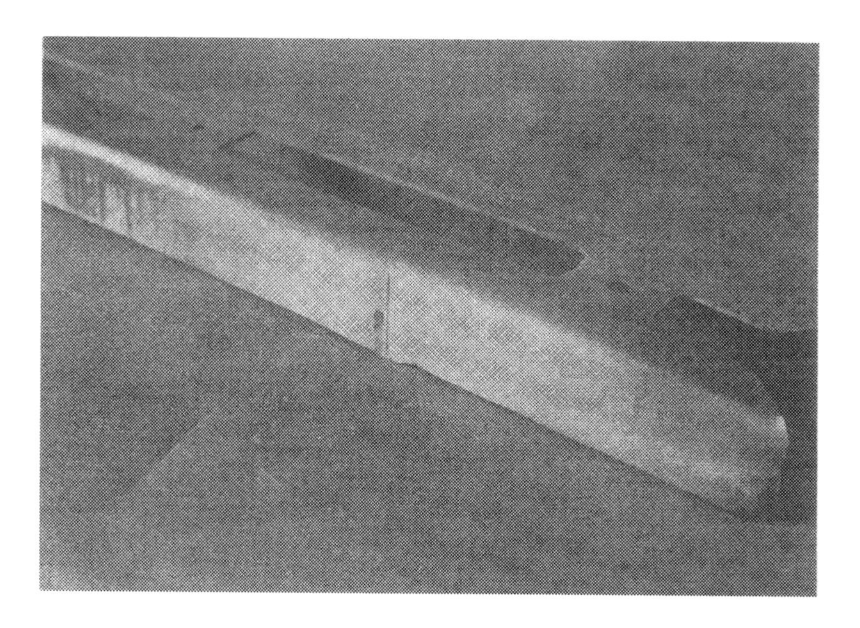

Fig. 2.7 A forming defect shown by a colour contrast penetrant process: the effect of incorrect forming of this aluminium strat are shown as two sharp parallel red lines where an incorrect forming procedure has led to folding of the material.

Fig 2.8 This small aluminium impeller shows an interesting inspection problem. The porosity which is shown throughout the component is not a reason for rejecting the component. However, in the vane at the bottom right a crack can be seen which goes through the thickness of the metal and is a serious defect.

Fig. 2.9 Red dye penetrant indications showing evidence of a porosity in an automative casting.

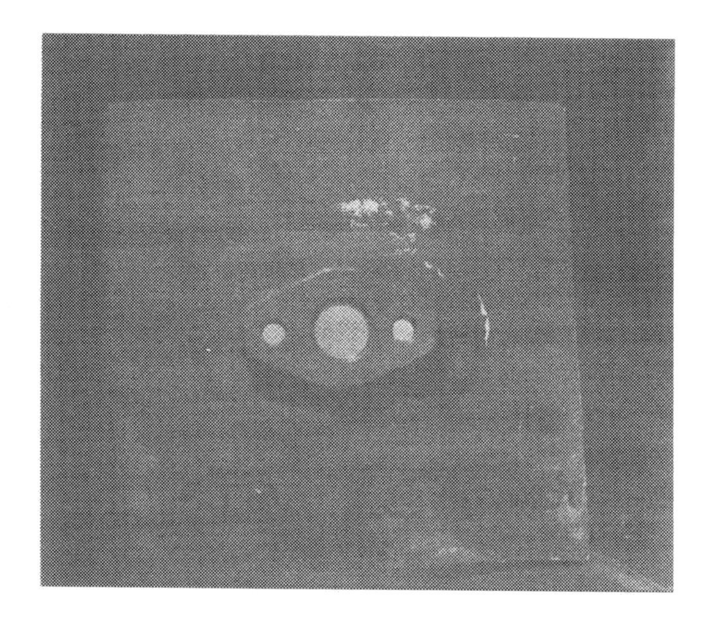

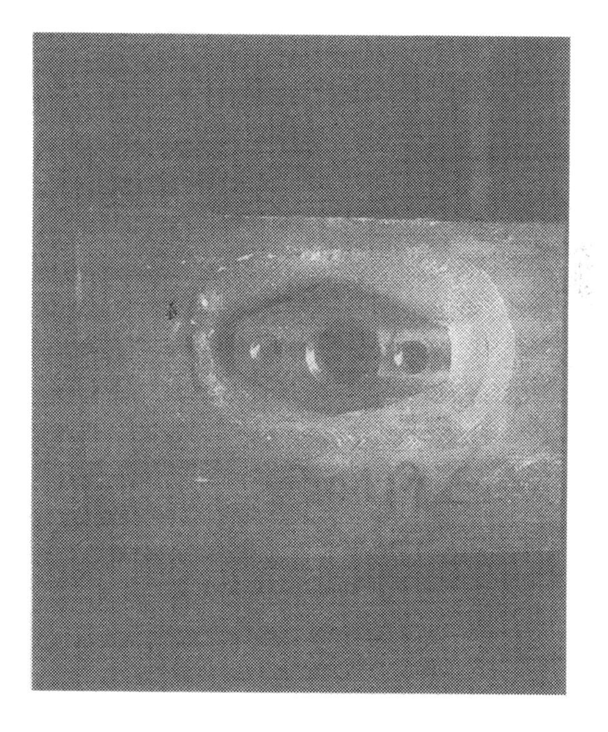

Fig. 2.10 Weld defects are difficult to identify from penetrant indications alone. These fluorescent indications of defects could be due to lack of fusion or cracking. The weld is not acceptable in either case.

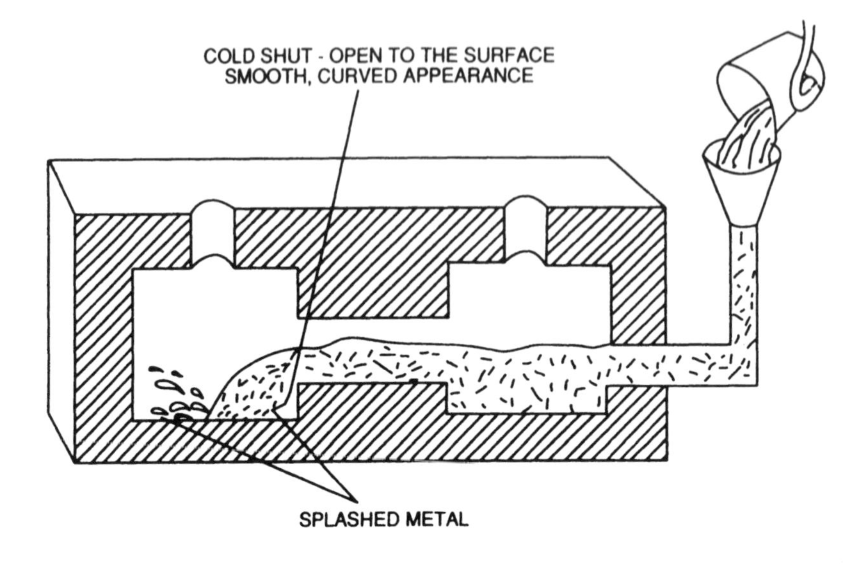

Fig. 2.11 Cold shut.

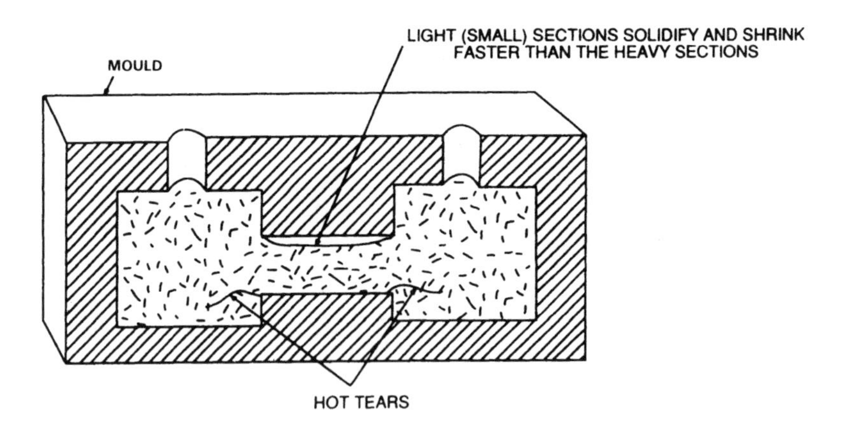

Fig. 2.12 Hot tears.

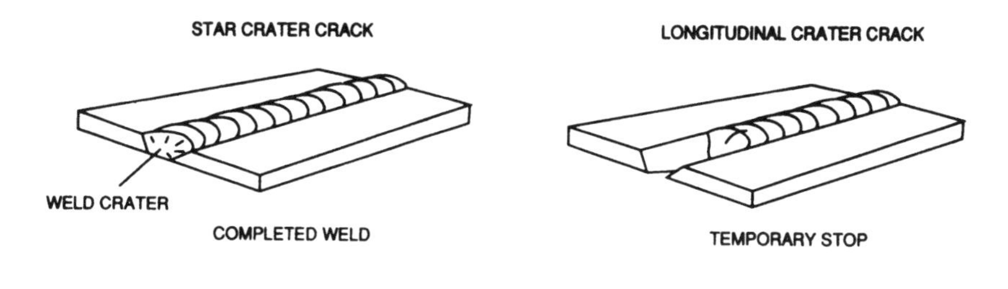

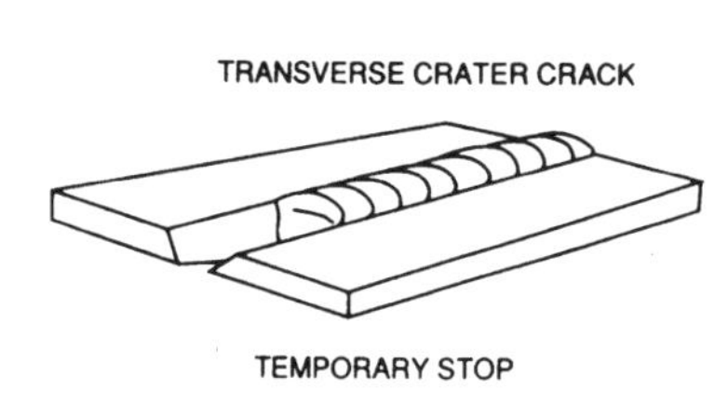

Fig. 2.13 Crater cracks in welds.

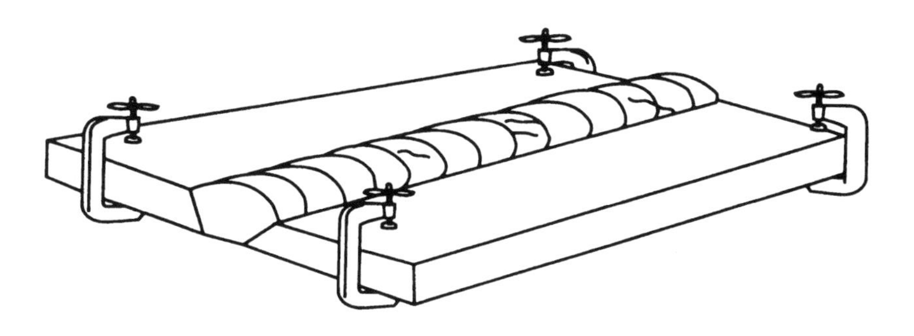

Fig. 2.14 Stress cracks in welds.

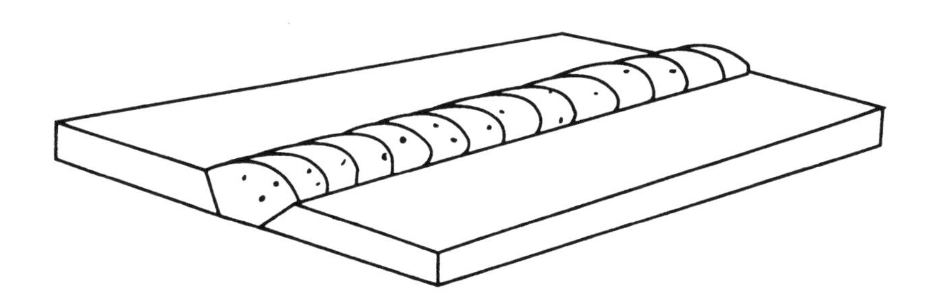

Fig 2.15 Porosity in welds.

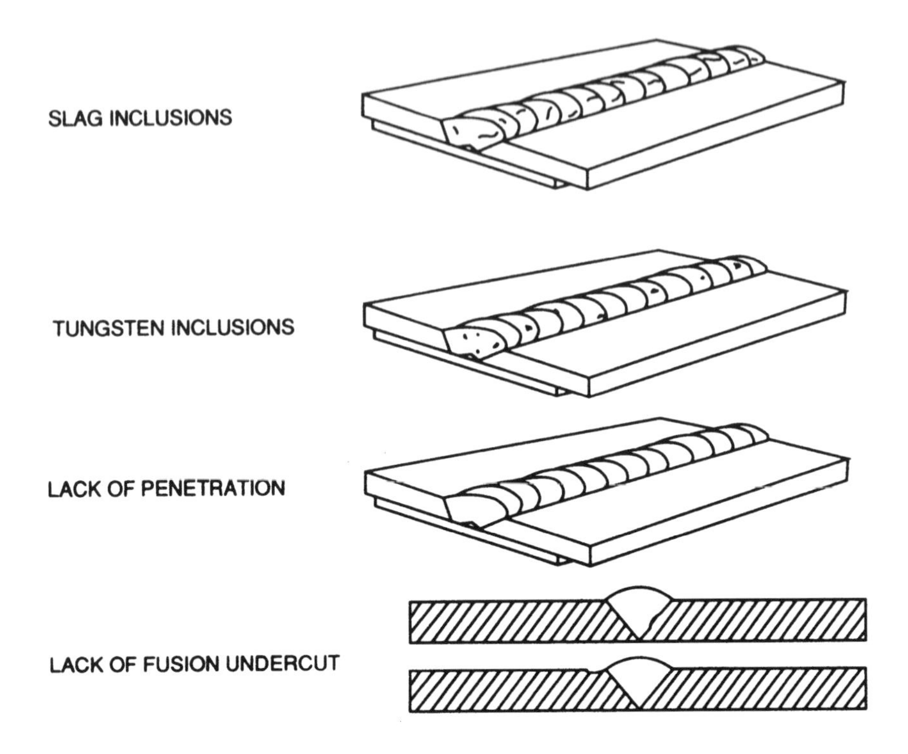

Fig. 2.16 Other welding discontinuities.

(k) Welding discontinuities

Many discontinuities arise during the welding process in either thin or heavy section (Figs 2.13–2.16). Lack of penetration, lack of fusion, undercutting, crater cracks, cracks in the heat affected zone and cracks in the weld metal are examples of discontinuities associated with welding. Another type of defect associated with welding is the appearance of non-metallic inclusions; these are also often called 'slag inclusions'. The process of welding uses new material to fill the space between two pieces of metal. During the welding process non-metallic materials can become incorporated in the weld material. Whether or not such inclusions are defects depends on the structure which is welded. They are certainly defects in critical rapidly moving parts, whereas in static structures they might be acceptable.

2.3.3 Secondary processing or finishing discontinuities

It is possible for all methods of processing and finishing of metals to produce discontinuities.

(a) Machining tears

When metals drag under a tool which is not cutting cleanly machining tears appear. Softer more ductile metals are more likely than hard metals to show this discontinuity.

(b) Heat treatment cracks

When metals are heated and then quenched in order to harden them cracking can occur if the operation is not carried out in a way which suits the material. Quenching cracks are often found where the material changes section or at fillets or notches. The margins of keyways, the roots of splines and threads are also likely locations for cracks. It is not only the cooling process which can cause cracking; too rapid heating can also produce this problem.

(c) Straightening cracks

Heat treatment can often cause warping or bending of parts owing to uneven cooling. Traditionally such deformation is corrected by straightening in a press. If the amount of bending required is too great or the part is very hard and brittle, cracking will occur.

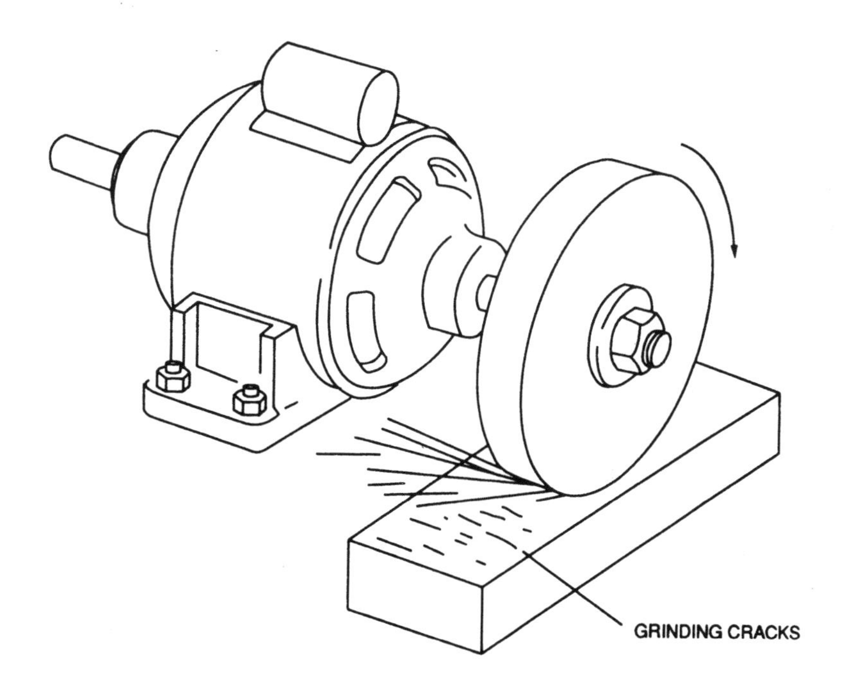

Fig 2.17 Grinding cracks.

(d) Grinding cracks

Surface cracking of hardened parts owing to improper grinding is a common problem (Fig. 2.17). These are a special form of thermal cracking and have noticeable relationship to quenching cracks. They usually occur immediately, but improper grinding can cause surface stresses and cracking which may not appear for some time.

(e) Etching and pickling cracks

Hardened surfaces which retain residual stresses may become cracked when pickled in acid. Acid attack on the surface allows the stress to become relieved by formation of a crack.

(f) Plating cracks

Apart from the chance that pickling prior to plating can cause cracks the actual plating process can cause problems itself.

2.3.4 Service cracks

The fourth class of discontinuity includes those which are produced during the working life of the part.

(a) Fatigue cracks

Fatigue cracking is a ubiquitous and serious problem particularly in metals which have to bear alternating dynamic stress or varying static stresses above the critical fatigue strength (far below the elastic strength). Fatigue cracks are frequently initiated by very small discontinuities at the surface (Figs 2.18 and 2.19).

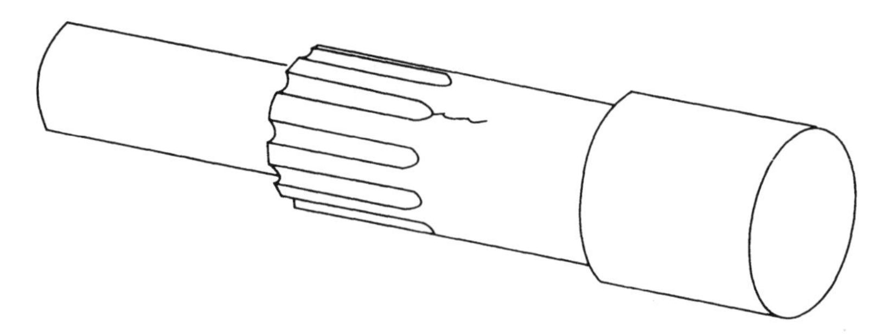

Fig. 2.18 Typical fatigue defect.

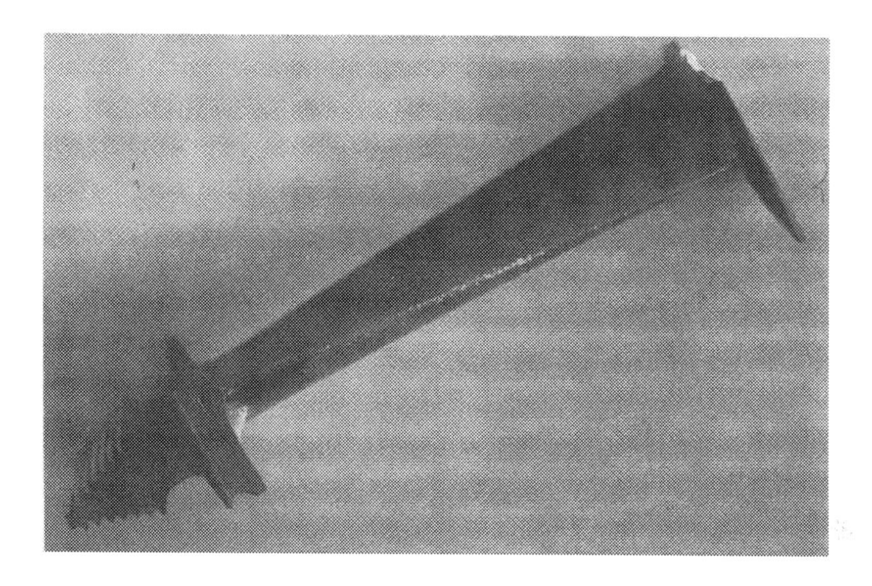

Fig. 2.19 Fluorescent indications of fatigue cracks on the leading edge of a turbine blade after penetrant processing.

(b) Pitting

When non-metallic inclusions remain in materials and the material is used to make moving parts such as roller bearings they can cause a special type of fatigue defect. In the inclusions close to the surface the action of the non-metallic material against the surrounding metal sets up stresses which eventually cause breaks to the surface which are seen as pitting. These defects will grow and eventually cause breakdown or non-function of the component.

(c) Corrosion

Parts which are held in tension and exposed to a corrosive environment in service can be expected to develop surface cracks which are termed stress corrosion cracks.

(d) Overstressing

Parts which in service are subject to stress above their design level are very likely to crack. Overstress can be caused by accident or unusual load, perhaps because of the failure of another member or component.

QUESTIONS

1. Failure of components may be due to

   (a) fatigue
   (b) overstressing
   (c) corrosion
   (d) all of these

2. Penetrant inspection can give indications of

   (a) non-metallic inclusions
   (b) sub-surface seams
   (c) forging bursts
   (d) segregation

3. Defects can be caused by any process which changes the shape of the material: true/false.
4. Porosity is a problem most associated with

   (a) forging
   (b) casting
   (c) extrusion
   (d) rolling

5. When inspecting components during overhaul, which of the following types of defect are most likely to be detected:

   (a) fatigue cracks
   (b) forging laps
   (c) grinding cracks
   (d) machining tears

Note: There is only *one* completely correct answer to questions 1–5. Tick your choice and check it with the correct answers on p. 211.

6. Define a defect.
7. Describe in a few sentences how dynamic loading is more likely than static loading to lead to fatigue failure of a component.
8. Explain what is meant by a stress raiser and give examples.
9. What are the four stages at which defects may be caused which are used for classification.
10. Give four examples of welding discontinuities.

3

Penetrant inspection: its development and application

A useful introduction to penetrant testing can be made by way of a historical survey of developments from the oil and whiting method used in the last century to the situation we have today.

The search for forbears of penetrant testing has included a number of test methods such as the use of soot on glazed china and other ceramics to find defects in the smooth surface. Despite obvious similarities such comparisons are largely retrospective. The oil and whiting method used by worried railway companies in the nineteeth century to detect cracks in axles of coaches and locomotives can be regarded as the true parent of modern penetrant testing. The oil and whiting process contained all the elements of modern penetrant testing. The black partially oxidized oil is the ancestor of the penetrant. The use of soapy water to remove surface excess has a very modern echo in the use of detergent removers, and the use of whitewash as a developer was an inspired choice - anyone who has been unlucky enough to paint over an oily spot, particularly with water-based emulsion paint, knows only too well how far that oily spot will spread and show through the finished paint.

After the introduction of magnetic particle inspection (MPI) this early penetrant process suffered a marked decline although its persistence or resurrection was such that there are NDT engineers still working who recall using this method in the 1950s. Generally, however, the oil and whiting method declined rapidly as MPI took root in NDT practice in the years from 1925 onwards. Most moving parts were then made of ferromagnetic steels and, despite the work of T. de Forest and others during the 1930s which produced quite sophisticated penetrant materials and techniques, MPI reigned supreme as the NDT method for showing surface and new surface defects. Effectively de Forest and other workers in the field at that time had a solution for which there was no problem - an academic curiosity.

The invasion of Poland in 1939 and the attack on Pearl Harbor directly changed this situation. Throughout history it has been an uncomfortable fact that periods of serious warfare have produced significant rapid development in other fields which have subsequently been of great benefit to the survivors and their descendants. During the Second World War great advances were made in the development of non-ferrous metals and in the uses of non-magnetic steels. The effects of the stimulus to investigate these metals is with us today, and directly or indirectly influences all our lives.

Clearly, non-ferromagnetic materials could not be tested for surface defects using MPI. Internal defects could be sought using X-ray methods that could be applied to iron and steel, but a new NDT method was needed to find surface defects. Here was a problem ready made for the penetrant chemists' answer. The potential was quickly recognized and in 1942 the first patent was taken out on a penetrant process. Somewhat curiously, this first patent was for a fluorescent penetrant. This fact is curious for two reasons: first, the model of the oil and whiting method used colour contrast of the black oil on the white paint where indications were viewed in daylight or under artificial white light; second, indications from fluorescent penetrant testing then, as now, required not only a UVA lamp (black light) to make the indications visible but also that viewing must take place in darkness. The restrictions that viewing in darkness placed on the process were quickly recognized, and the introduction of colour contrast penetrant materials (red) followed swiftly.

At this stage all the penetrants were formulated to be water washable. From the beginning it was known that hydrocarbon oils have excellent penetrating properties; the oil and whiting method had shown this. However, oil does not mix with water, and whereas solvents did exist which could remove surface excess penetrant they were then, as now, difficult to use for anything other than local inspection of a small area. This was not very helpful for inspection of large parts or large numbers of parts. A water-washable penetrant allowed easier and uniform removal of surface excess penetrant. The chemical problem of making oil mix with water was overcome by using an emulsifying agent. A simplified view of the formulation of a water-washable penetrant in 1942 is of oil plus an emulsifying agent and a dye, either red or fluorescent. The actual chemistry was more complicated even then, but the model is useful.

In those early days water was considered as the universal remover. Developers, however, showed more diversity. In some instances none were used; the oil and whiting method left its legacy of an inert white pigment suspended in water. This was particularly useful for red dye colour contrast penetrants where, unless the material under test has a pale surface, e.g. clean aluminium, viewing the indications is very difficult or even impossible without a contrasting developer. Dry powders were introduced very quickly, as were non-aqueous suspension developers. Water is fairly volatile but it does not dry quickly enough to work satisfactorily as a carrier for penetrant developers unless it is actively heated. This was a distinct disavantage when penetrant testing was carried out on site. The use of a volatile solvent overcame this problem. It is likely that some unpleasant chemical such as carbon tetrachloride was used in the early days, as this and other solvents had not then acquired their sinister reputation. Carbon tetrachloride and related chemicals have an evaporation rate which is quite good for this purpose and it has the advantage of being non-flammable. Over a period of time different volatile halogenated hydrocarbons were used with progressively less unpleasant potential health hazard until 1,1,1. trichloroethane became established. The association of this chemical with depletion of the earth's ozone layer has led to its being banned since the mid 1990s. Since then flammable volatile organic solvents such as alcohols, esters, and ketones have been used. There is pressure to reduce or eliminate the use of all volatile organic compounds (VOCs) and this solution may be temporary. It will be interesting to see how any such change is accommodated by penetrant testing.

In the years 1942-8 penetrant testing could be carried out in a number of ways. The available materials are summarized in Table 3.1. Solvent is included in this table as a remover since it is very unlikely that the advantage of a volatile solvent over water in terms of drying time had not been discovered. Little thought is often given to the proper use of solvent remover nowadays, and so it is unreasonable to expect more care to have been taken in the 1940s.

Table 3.1 Materials available for penetrant testing between 1942 and 1950

Penetrants	*Removers*	*Developers*
Colour contrast	Water	None
		Aqueous suspensions
Fluorescent	Solvents	Non-aqueous suspensions
		Dry powder

At this stage there were at least 16 ways of carrying out penetrant testing with the limited choice of materials then available. As is normally observed throughout history, wartime had led to very significant advances in technology and science. In 1945 and 1946 the atom was split in a devastating way twice over Japan and in a less devastating way on a number of occasions elsewhere. The possibility of using this power in a constructive way was not lost on the world. German rocket research had advanced to a stage where London could be bombarded from continental Europe. As well as advancing material science itself, this development accelerated the development of jet engines to propel aircraft fast enough to chase these rockets and destroy them in the air before they reached their targets.

After the conflict was settled the progress in engineering, metallurgy, physics, chemistry and other branches of applied and theoretical technologies was not wasted but turned to peaceful use as well as preparation for possible future military use. Materials science had developed very rapidly, opening the way to such spectacular developments as visiting the Moon, landing on neighbouring planets etc., and more mundane but no less spectacular achievements as flying up to 400 people half-way round the world in a relatively short time.

These advances demanded that materials work harder and in much more aggressive environments. Materials such as titanium alloys, which previously had been found almost exclusively in the laboratory, had become important engineering materials. Ferromagnetic materials were in retreat, albeit a limited one, as aluminium became a significant material.

All these advances, uses of new materials in engineering and increased demand on materials had their influence on penetrant testing. Users of penetrant testing started to ask for a system which could be relied upon to show the shallow cracks initiating fatigue cracking at an early stage. Water-washable penetrant penetrates such discontinuities satisfactorily enough but is also readily washed out again. It was more than 30 years before a water-washable penetrant which could be used for detecting such defects satisfactorily became available, and even now it is only used when the shape and conformation of a part demand it.

The answer provided by the penetrant chemists in the late 1940s was to separate the emulsifier from the penetrant and apply it after penetrant application in a separate stage. This had two immediate advantages. First, it was possible to formulate a penetrant which actually resisted water and, second, it was possible to dissolve more coloured or fluorescent dye in such a penetrant. Emulsifying agents in general and members of the sulphonate family which were used in early water-washable penetrants are not good solvents. Dye concentration is not the only factor in potential sensitivity of penetrant systems but it is one of the four most important, and at that stage in the history of penetrant testing increase in dye concentration allowed a significant advance in performance.

The system using a non-water-washable penetrant followed by application of an emulsifier became known as the post-emulsifiable penetrant process. It was immediately successful in improving the performance in finding small shallow cracks.

During the following 10 years both water-washable and post-emulsifiable penetrant materials were developed so that more sensitivity could be obtained, and in the case of fluorescent penetrants at least two levels of potential sensitivity became available in both systems and research into improving emulsifiers had led to the availability of two types – rapid and slow.

During the 1950s the Rolls Royce company carried out work to improve the removal process for non-water-washable penetrant. Their concern was that components which were processed by using the emulsifying agents then available often showed quite irregular quality of background fluorescence or colour. On vertical surfaces excellent background was achieved, whereas on horizontal surfaces heavy residual fluorescence or colour could obscure indications of defects. Therefore contact times with the emulsifying agent were governed by the time needed to give good background on horizontal surfaces. The NDT engineers at Rolls Royce feared, quite correctly, that such times were too long for the vertical surfaces and that these would suffer over-removal and indications of defects would be lost. A major reason for the difference in the results of the process is that after immersion in penetrant and drainage the film thickness of the penetrant on the part varies from being very thin on vertical surfaces, where it drains readily, to very thick on horizontal surfaces, where it remains undisturbed or even collects. The result of this is that the emulsifying agent has more penetrant to deal with on some parts of a component compared with others, and hence there is uneven removal of the surface excess penetrant and very variable background for the inspector. The engineers at Rolls Royce sought ways of creating a thin uniform film of penetrant on all surfaces of the component irrespective of whether they were vertical or horizontal, and this was achieved by using vigorous water spray which physically removed most of the surface excess penetrant leaving a thin film on all areas. A problem arose in that this prerinse, as it is now known, also left a wet component and excess water interferes with emulsifying agents. This led to the use of liquid soaps and detergents which tolerate water. The proper formulation of these materials and the effects of their concentrations on the process were studied and thus a new process was born.

Because the new removers were used as solutions in water they were termed 'hydrophilic'. The old oil-based emulsifiers which were used as supplied were termed 'lipophilic' to distinguish them. Actually the difference between the actions of these two removers is quite fundamental and it is becoming more common to hear detergent solutions referred to as hydrophilic removers which is much more satisfactory.

True emulsifiers mix with the surface excess penetrant and make it water washable. This is why emulsifier contact time is so critical and must be controlled closely. If it is not, the emulsifier will attack penetrant trapped in defects, making it water washable, and since it is in infinite concentration it will create a mixture which is easier to wash out than a properly formulated water-washable penetrant. A hydrophilic remover is used as a solution in water. It works by changing the surface free energy between the penetrant and the solid surface so that the contact angle increases from near zero, with good wetting, to around 180° and the penetrant stands up in globules much as rainwater stands up on the bodywork of a well-polished car. These globules separate from the surface and become stabilized in the liquid.

The late 1960s saw two further developments in the introduction of water-soluble developers and dual-purpose penetrants. Water-soluble developers are blends of salts which dissolve in water to give clear solutions. They are applied by immersion, wet spray, flow-on or brush. As with the aqueous suspensions these developers must be dried after application. Unlike the suspensions they do not need continuous agitation during application and in many cases have replaced aqueous suspension developers. Dual-purpose penetrants work both as a normal colour contrast penetrant and when inspected under UVA (black light) illumination fluoresce. Available products fluoresce pink or orange. The reason for having these materials is to allow the inspection of hollow items such as large tanks or gun barrels. They are dark inside and the penetrant indications can be inspected using a UVA lamp (black light) with a maximum wavelength at 365 nm. Outside the same item the penetrant system can be used in the colour contrast mode.

By 1970 the usual materials now used in penetrant testing had been developed. Since that date the history of penetrant testing has been concerned with refinement and improvement of both materials and the techniques used to apply them. Table 3.2 summarizes the development of penetrant testing.

Table 3.2 Chronology of development of penetrant testing

Date	*Development*
1880–1920	Oil and whiting method in widespread use on railway axles and boiler plate
1935–1940	Betz, Doane and de Forest work on penetrant techniques
1941	First patent on penetrant test materials
1945	Colour contrast penetrant system developed by Rebecca Sparling
1948–1955	Introduction of post-emulsifiable penetrant system
1957	Hydrophilic penetrant removers patented by Rolls Royce
1969	Dual-purpose penetrant system introduced
1978	Special penetrants for use with hydrophilic removers introduced
1980–1985	Special low tempterature (to −30 °C) and high temperature (to 180 °C) systems introduced
	Heat-fade-resistant fluroescent penetrants introduced

Refinement of the materials has included the following.

(a) Designing penetrants specifically to be used with hydrophilic remover solutions

When hydrophilic removers were initially introduced they were simply applied to the existing post-emulsifiable penetrants with, it must be said, great success. However, it is possible to design penetrants whose characteristics are more favourable to the hydrophilic removers. Research in this area has given us penetrant systems which provide excellent sharp indications against a clean background, thus making inspection easier. This has been achieved without loss of sensitivity.

(b) Heat-fade-resistant fluorescent penetrants

The volume of penetrant in a defect 3 mm long, 50μm deep and 2 μm wide is so small that it is little short of miraculous that the process works at all in a practical way. The effect of overheating this tiny volume of liquid either by excessive temperature or by leaving parts too long in a dryer at a a moderate temperature can lead to severe fading of the fluorescent brilliance of the indication. Careful research has led to the development of penetrants which give indications which fade by less than 6% when parts are heated to 100 °C for 30 min.

(c) Materials for use at high temperature

Penetrant testing at temperatures above 80–100 °C must involve colour contrast penetrants. Fluorescence is adversely affected by heat, and the fluorescent brilliance of these penetrants will fade as the temperature rises, falling to zero at around 220 °C. The process is reversible, and on cooling the original characteristics return provided that care has been taken not to lose volatile solvents in the process. Some penetrants retain their fluorescence when used at temperatures up to 80 °C. Modern colour contrast systems exist which can be used at temperatures up to 200 °C. These are very useful when welds are inspected as the time lost in waiting for the part temperature to fall to 200 °C is much less than that lost in waiting for the part to reach 40, 50 or even 60 °C.

(d) Penetrant materials for use at low temperature

Penetrant testing depends on physical phenomena, most of which can be expected to change by a factor of 50% for every 10 °C change in temperature. On this basis a penetrant which works in 10 min at 20 °C will take 20 min at 10 °C and 40 min at 0 °C. At − 30 °C the same phenomenon will take 320 min, which is tantamount to saying that it will not happen at all since at these temperatures the phenomena are so complex that the linear relationship breaks down. The behaviour of materials at low temperatures is itself very different from that at ambient temperatures of around 20 °C, and as man has invaded hostile areas of the world such as the Arctic in the search for oil or siting of military bases so the need to carry out NDT and in particular penetrant testing has followed. To meet this need penetrant testing processes, both fluorescent and colour contrast, have been developed. The only concession that they make to the cold is that the developer requires 30 min as opposed to the usual 5 or 10 min. The laws of physics can be stretched but not broken.

(e) High resolution penetrant systems

Although high resolution penetrant systems pre-date 1970, they are dealt with here as special systems. Unfortunately, in practice they can only be used as manual systems. Special solvents are used for the penetrants to carry the dyes into defects. Equally special techniques are used to remove the surface excess, and then a lacquer developer is applied. Lacquer developers are closely related to non-aqueous suspensions but they have a resin dissolved in the carrier solvent. As the solvent evaporates and the indications develop the resin comes out of solution and simultaneously freezes the indications. If care is taken it is possible to achieve very high levels of sensitivity and excellent resolution.

(f) Developers

Early developers, whether dry powder, aqueous suspension or non-aqueous suspension, were crude. Until recently some users still refer to non-aqueous suspensions a chalk in 1,1,1-trichlorethane (more likely 'trike'), which is enough to make any penetrant chemist wince mentally, but allowing for chemical errors early non-aqueous suspension developers were as crude as that. All types of developer have undergrone continuous improvement which is still under way. It must be appreciated that developers do have their own sensitivity contribution to the penetrant process and it is dangerous to ignore or forget this.

CONCLUSION

Sadly, penetrant testing is somewhat of a Cinderella of NDT. The analogy is excellent. Cinderella was the most useful member of her household and the least appreciated – just like penetrant testing. Penetrant testing suffers from the simplicity with which it can be applied successfully. There are no magic black boxes understood only by serious-looking men in white coats. The results are not easily recognized and analysed by machines – an unforgivable sin in our current times. It can be done but is costly and difficult; the fault is that human eyes are still well ahead of machines in most phases of image recognition. Despite these disadvantages, penetrant testing is thriving. It is fairly inexpensive, and when carried out by dedicated operators it enjoys surprisingly high levels of confidence. The actual processing (if not inspection) can be automated successfully. It can be used for inspection for surface defects on any non-porous surface provided that the chemicals do not react with the surface. It is probably the least shape sensitive of the available NDT methods for surface inspection. When an engineer designs something he or she is not thinking of NDT problems but function, and some of the results are truly horrifying to the NDT staff. Penetrant methods can be designed to inspect the most bizarre of shapes. Size is not a real problem either, as both vast areas of surface and tiny components can be inspected.

A recent estimate showed that over 80% of a jet engine is inspected by penetrant at some time during manufacture. Many other items on an aircraft, such as structural frame parts, landing gear and wheels, are also penetrant tested both during manufacture and throughout service life. Penetrant testing is also used in many other industries. Gas turbines used in the petrochemical industry and in power generation, components for nuclear power generation, non-magnetic automotive components and even surgical replacement joints are examples where penetrant testing is used as part of the quality control programme.

Recent research has indicated that the use of solvent removers in the vapour phase or immersion in the liquid phase with ultrasonic agitation may become commonplace for the inspection of ceramic and powdered metal coatings. These coatings are now quite commonly used on aero-engine components to protect the basic metals from excessive heat. Cracks and other discontinuities in these coatings are undesirable and are often extremely difficult to find. Conventional penetrant testing is not satisfactory as a heavy background fluorescence is left which masks any indications.

Two approaches have shown that penetrant inspection can be most useful if special techniques are used. The first successful special technique is to use a penetrant which consists essentially of fluorescent dyes in a volatile solvent system. After application the penetrant is allowed to dry out on the surfaces under inspection and then passed to a removal process which uses an organic solvent in the vapour phase for up to 3 min or to an ultrasonically agitated bath of solvent where the components are immersed for times up to 4 min. The second approach which has been successful on these surfaces is to use a conventional penetrant which can be diluted with some colourless liquid to reduce the fluorescent brilliance of the material. In this case the diluted penetrant is applied by any of the accepted methods and then subjected to one of the removal methods described above, i.e. vapour phase solvent removal or removal in a bath of ultrasonically agitated solvent.

It is interesting to note that in both cases the indications appear black against a fluorescent yellow-green halo. Some indications appear sharp black with vivid fluorescent yellow-green areas around the indication where penetrant bleeds out. This is due to self-quenching of the fluorescence of the material in the defect. In the second case the indications appear fluorescent yellow-green, as is the case with conventional penetrant testing when the dilution of the fluorescence is very high. However, early results suggest that some sensitivity is lost before the transition from quenched black to conventionally fluorescent indications.

QUESTIONS

1. The first commercially available penetrants were

   (a) red and solvent removable
   (b) fluorescent and post-emulsifiable
   (c) red and water washable
   (d) none of these

2. Introduction of the post-emulsifiable penetrant process allowed the following type of defect to be detected:

   (a) deep fine cracks
   (b) shallow cracks typical of the early stages of fatigue
   (c) porosity
   (d) cold shuts

3. Introduction of the detergent removers (hydrophilic emulsifiers) was the result of

    (a) the need to provide a uniform layer of penetrant over the complete surface of components
    (b) the need to reduce cost
    (c) the need to ensure that a better level of background fluorescence was achieved
    (d) problems with post-cleaning of titanium components

4. Dual-purpose penetrants

    (a) give indications of both subsurface defects and those open to the surface
    (b) give indications which can be inspected under white light and will fluorescence under UVA (black light)
    (c) were developed to allow penetrant processing to be used on non-metallic materials
    (d) were developed to allow penetrant to be applied by both immersion and electrostatic spray

5. Modern penetrant developers may be

    (a) dry powers
    (b) non-aqueous suspensions
    (c) aqueous solutions
    (d) aqueous suspensions
    (e) all of these

Note: There is only one correct answer to questions 1–5. Tick your choice and check it with the correct answers on p. 211.

6. Describe how the oil and whiting methods for detecting cracks shows itself to be the true forerunner of modern penetrant testing.
7. Explain two advantages which were results of the introduction of the post-emulsifiable pentrant process.
8. What advantages might water-soluble developers have over water-suspendable developers?

4
The principles of the processes

Liquid penetrant testing is a non-destructive method for locating discontinuities which are open to the surface of essentially non-porous solids.

4.1 THE PROCESS IN OUTLINE

Despite the fact that penetrant testing has been exposed to the ingenuity of engineers, technologists and scientists for over 40 years, the outline remains simple and unchanged. The task is to get the penetrant liquid with its coloured or fluorescent dye into defects. The surface excess must be removed sufficiently well to allow good contrast between any indications and the background and the parts prepared for inspection. This last step normally involves use of a developer. Figure 4.1 illustrates the basic steps diagrammatically. The individual steps are discussed in detail in the remainder of this chapter.

An important factor which must be remembered in *any* phase of penetrant testing is that the whole system must be considered in its entirety. Although each and every step must be considered and studied individually it is the complete scheme which must be considered. The point will be dealt with more thoroughly in Chapter 6, but must be remembered at all stages. First, we start with a part or possibly a raw material which needs to be inspected using a liquid penetrant system. Based on a number of considerations we shall choose the following:

1. fluorescent or colour contrast system
2. mode of removal of surface excess penetrant
3. type of developer

It is essential that we remember that we are testing parts or materials rather than applying chemicals to surfaces. A further interesting thought is that we do not carry out penetrant testing or any other form of NDT in order to find cracks or defects but to find parts or materials which are not defective and are fit for their purpose.

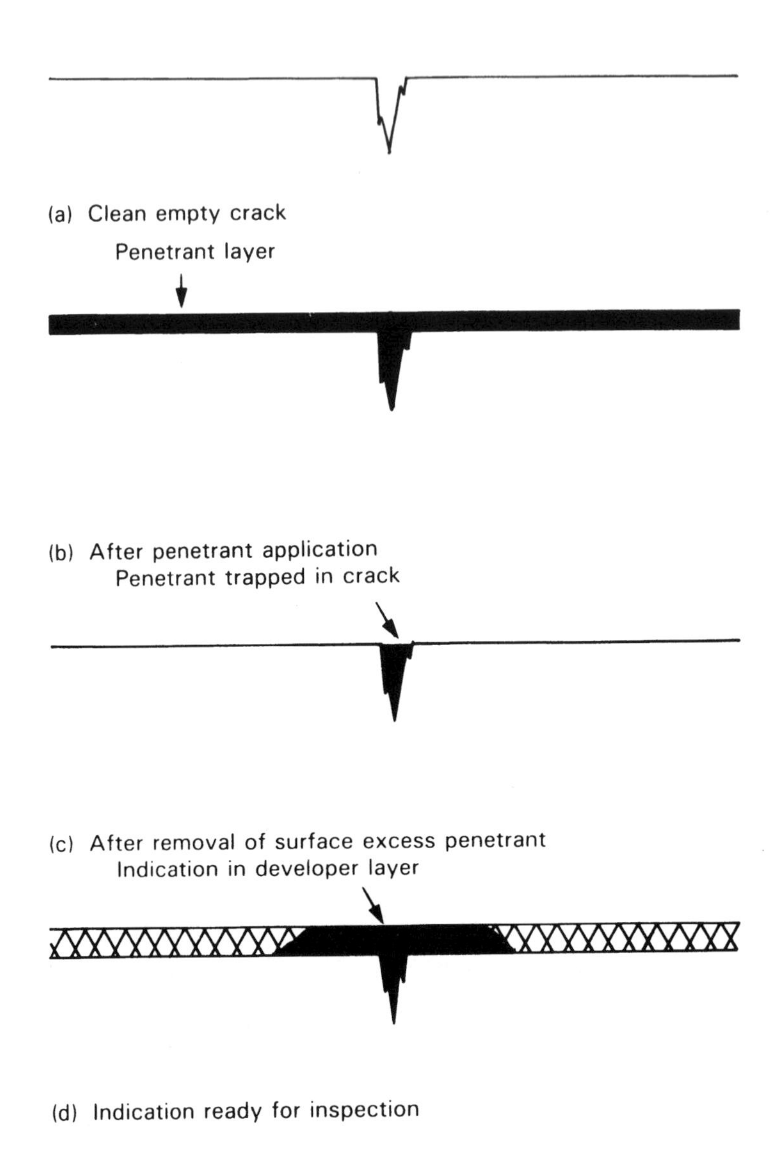

Fig. 4.1 The basic steps of penetrant testing.

4.2 THE PHYSICS AND CHEMISTRY OF THE PROCESS

4.2.1 How penetrants work

The way in which penetrants work is frequently explained in terms of a simple capillary rise phenomenon. This explanation suggests that the liquid enters cracks, porosity, laps etc. by the same mechanism as liquid rises in a tube. Figure 4.2 illustrates the basis of this theory. The simplified model suggests that a crack or other discontinuity resembles a tube with a closed end and the penetrant enters by a capillary rise type action. As is well known the liquid will rise most in the narrowest tube and the rise decreases as the diameter of the tube increases. This is described by the formula for capillary pressure in a vertical tube:

$$P = \frac{2S \cos \theta}{r}$$

where P is the capillary pressure, S is the surface tension of the liquid, θ is the equilibrium contact angle of the liquid with the solid surface and r is the radius of the tube. In the case of a crack r is taken to represent the width of the crack. From this equation it is seen that as r increases so P decreases, and this agrees with the observed phenomenon (Fig. 4.2). This approach suggests that penetrant has a much stronger tendency to invade tight fine discontinuities than broader ones. Experience shows this to be untrue, and so there is a basic objection to this theory which cannot be resolved.

The above explanation is quite unsatisfactory and we prefer the explanation that penetrant enters by reason of forces set up by the immersion free energy of the solid surface. All surfaces have a free energy associated with

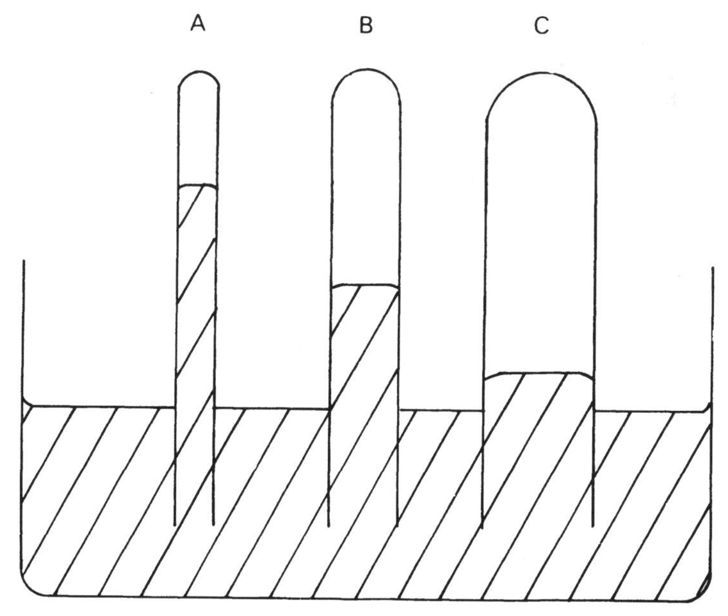

Fig 4.2 Diagram of capillary action.

them, otherwise they would not exist. In liquids this is recognized as surface tension. It can be calculated that liquids wet surfaces and invade discontinuities because of the physical reaction resulting from the free energies of the surfaces of both the liquid and the solid. The free energy of any surface involves at least two substances unless the measurement is made in a vacuum. If we consider a piece of metal the surface free energy is between the metal and air. If we then immerse the metal in a liquid, the surface free energy will change and the relationship will be specific to the particular combination of metal and liquid, varying only with changes in temperature, volume and pressure although the pressure variations in liquid systems are almost negligible.

Complete experimental evidence that this is the way that penetrants invade surface defects and other discontinuities is not yet available. However, the theoretical model is good, and when worked out has the advantage of suggesting that tight fine defects are more difficult to penetrate than wide ones, and this concurs with more than 45 years of experience.

Where capillary action is very important is in the action of penetrant developers, which is dealt with later.

4.2.2. Penetrant removal

The whole process of liquid penetrant testing is most vulnerable during the removal of surface excess penetrant. When we consider what we want to achieve at this stage this vulnerability is not surprising. Removal of surface excess penetrant requires fairly aggressive physical and chemical action sufficient to remove highly coloured liquids from close contact with the solid surface. At the same time this action must not be so aggressive as to remove penetrant from entrapment in discontinuities. The ultimate challenge to penetrant system chemists is to devise a system which will show small surface defects in a very porous material.

Three main types of action are used for removal of surface excess penetrant:

1. emulsification
2. detergency
3. solvent action

Emulsification is involved when a penetrant is water washable or if lipophilic emulsifiers are used with a non-water-washable penetrant, detergency is the phenomenon acting when solutions of hydrophilic removers are used, as in the modern post-removable process, and solvent action occurs when an organic solvent is used to remove surface excess penetrant.

(a) Emulsification

Emulsification is the process whereby oils are made miscible with water. Emulsifying agents fall into four groups:

1. anionic
2. cationic
3. amphoteric
4. non-ionic

The first three groups involve the formation of electrically charged molecules when the emulsifying agent is mixed with water. Early water-washable penetrants used ionic emulsifying agents. However, modern materials do not. Ionic species can cause changes in pH values (acidity and alkalinity) when water is introduced and can thus be unsuitable for metallurgical reasons. Modern water-washable penetrants and lipophilic emulsifiers use the fourth group of non-ionic emulsifying agents. These chemicals can be thought of as being made up of long chain-shaped molecules, one part of which is attracted to oil and the other to water thus forming a bridge between these two types of chemical which normally do not mix together.

Water-washable penetrants can be considered as a mixture of a solution of coloured or fluorescent dye in oil with an emulsifying agent. The truth is more complex than this, and many modern water-washable penetrants contain up to 15 different chemicals rather than three, but the principle still holds. A similar situation occurs when a lipophilic emulsifier is used to remove surface excess non-water-washable penetrant.

In the second case the penetrant is often strongly hydrophobic and resists mixing with water very strongly. After the penetrant has been applied and left to penetrate discontinuities the emulsifier is applied and left on the part long enough to make the surface excess water washable while leaving the penetrant entrapped in discontinuities unchanged. This process clearly relies on timing and, to some extent, skill in handling the parts and chemicals. When water is added to this essentially two-liquid mixture a number of physical changes occur allowing the penetrant–emulsifier mixture, whether designed as a water-washable penetrant or occurring in a post-emulsifiable process, to be washed away leaving entrapped penetrant *in situ*.

The danger is over-removal either by too vigorous a water wash in the case of water-washable penetrant (this was the underlying reason for introduction of the post-emulsified process) or too long a contact with the emulsifier, thus allowing penetrant entrapped in discontinuities to become water washable. Despite these potential hazards water-washable penetrants have been formulated satisfactorily and are part of a wide range of very satisfactory penetrant testing systems, with five levels of sensitivity from colour contrast and four levels of sensitivity using fluorescent materials.

(b) Detergency

The basis of detergency is soap, one of those happy accidents of invention which, in principle, has never been improved upon. The success of the original soaps, which were and are mixtures of the sodium salts of fatty acids, became apparent as soon as attempts were made to improve on the characteristics of the original. Soaps are anionic detergents where detergent removers are non-ionic. However, the physical principles are the same.

All liquids and all solids have surfaces. If this were not so all matter would be mixed together in a continuum much in the way matter is believed to coalesce in a forming star. Associated with all surfaces is energy providing the force to hold them together, and this is termed the surface free energy.

The surface free energy of a solid or liquid varies considerably according to what it is in contact with. The surface free energy of an oil droplet in air is quite different from that of a similar droplet in water. Similarly, the surface free energy of a solid surface varies markedly depending on whether it is exposed to the air, covered with oil or in contact with a detergent solution. In fact we cannot consider surface free energy to be a fixed value at all but a variable property depending on the number and characteristics of the different compounds present and other considerations such as temperature and pressure. Detergency takes full advantage of this.

A good indicator of the surface free energy of a system is the contact angle between a liquid and the solid surface on which it lies. When magnesium or aluminium is coated with an oil, e.g. a penetrant, and the rest of the system is air the contact angle is very small indeed and we say that it wets well. Stainless steel has a different surface free energy and some penetrants will retract from the edges of the part or changes in section. On tungsten wetting can be a serious problem. When the air in such a system is replaced by remover solution the whole system changes and the oil does not wet at all. The contact angle between the penetrant and the solid surface rises from near zero to near 180 °, and a globule of penetrant stands up like rainwater on a well-waxed car body and becomes separated from the surface.

The complete picture is more complicated than this and at temperature above 50 °C a second more aggressive mechanism takes over. However, since in penetrant testing hydrophilic removers (detergent solutions) are used at temperatures below 40 °C this can be ignored.

There is a second phase of detergency which has led to such confusion with regard to this process that some specifications and textbooks consider detergent removers as some kind of 'water-based emulsifiers'. This is not the case, and the difference must be recognized in order to appreciate the real value of the post-removable penetrant testing process.

The confusing step is that, after being separated from the solid surface, the oil gobules become stabilized in the bulk of the remover solution. This is achieved by the formation of micelles. When detergents are dissolved in water they show the apperance of true solutions. However, the molecules are large and have oil-favouring and water-favouring parts just like true emulsifiers. The oil-favouring sites aggregate together centrally in spherical forms, exposing the water-favouring parts to the surrounding water. These aggregates are called micelles and, in the absence of oils, are short lived, being continuously broken up and re-formed. However, when an oil globule becomes available it is seized, stabilized in the centre of the micelle and floated off into the water. The clue to the marked difference between detergency and emulsification is that, whereas emulsifying agents have an infinite tolerance to penetrants, remover solutions have a somewhat low tolerance to them. A 10% solution of hydrophilic remover has a tolerance to penetrant of well under 10%.

(c) Solvent action

Solvent action is the most aggressive of the removal processes. The formation of a solution is one of the more powerful phenomena in physical chemistry. A solution can occur in any physical phase - liquid, solid or gas. In all phases it is a complicated process which has been the source of many studies and a number of important physical chemical laws. In penetrant testing we are concerned only with liquid solutions. When a solution is formed the mixture is so intimate that it cannot be separated by any physical method. This is quite distinct from a suspension, a colloid or an emulsion.

Penetrants are themselves solutions, and in penetrant chemistry a solvent is any chemical which will dissolve the whole mixture. To be useful such solvents must be readily volatile at temperatures between 15 and 30 °C; otherwise they offer no advantage over water. A solution can be regarded as a mixture of molecules where all types of molecule are essentially equivalent and so form a continuous homogenous lattice. Clearly, any chemical which can enter into such a system with the penetrant can be very difficult to control. If the solvent is applied in such a way that it is in infinite concentration with respect to the penetrant it will act to remove penetrant not just from the surface but also from within discontinuities. It is for this reason that solvents may not be applied by spray, immersion or flooding. All these methods lead to an infinite concentration of solvent with respect to the penetrant residues in discontinuities. Applications using any of these methods causes marked loss of sensitivity.

4.2.3. Developer action

There are three types of penetrant developer in common use today: dry powders, non-aqueous suspensions and aqueous solutions. One further type, the aqueous suspension, is still used but has largely been replaced by aqueous solutions. There is also the specialized family of lacquer developers. In addition, there is the possibility of using no developer at all.

The value of using developers as part of a penetrant process is illustrated in Figure 4.3. These three photographs show a crack in a test panel

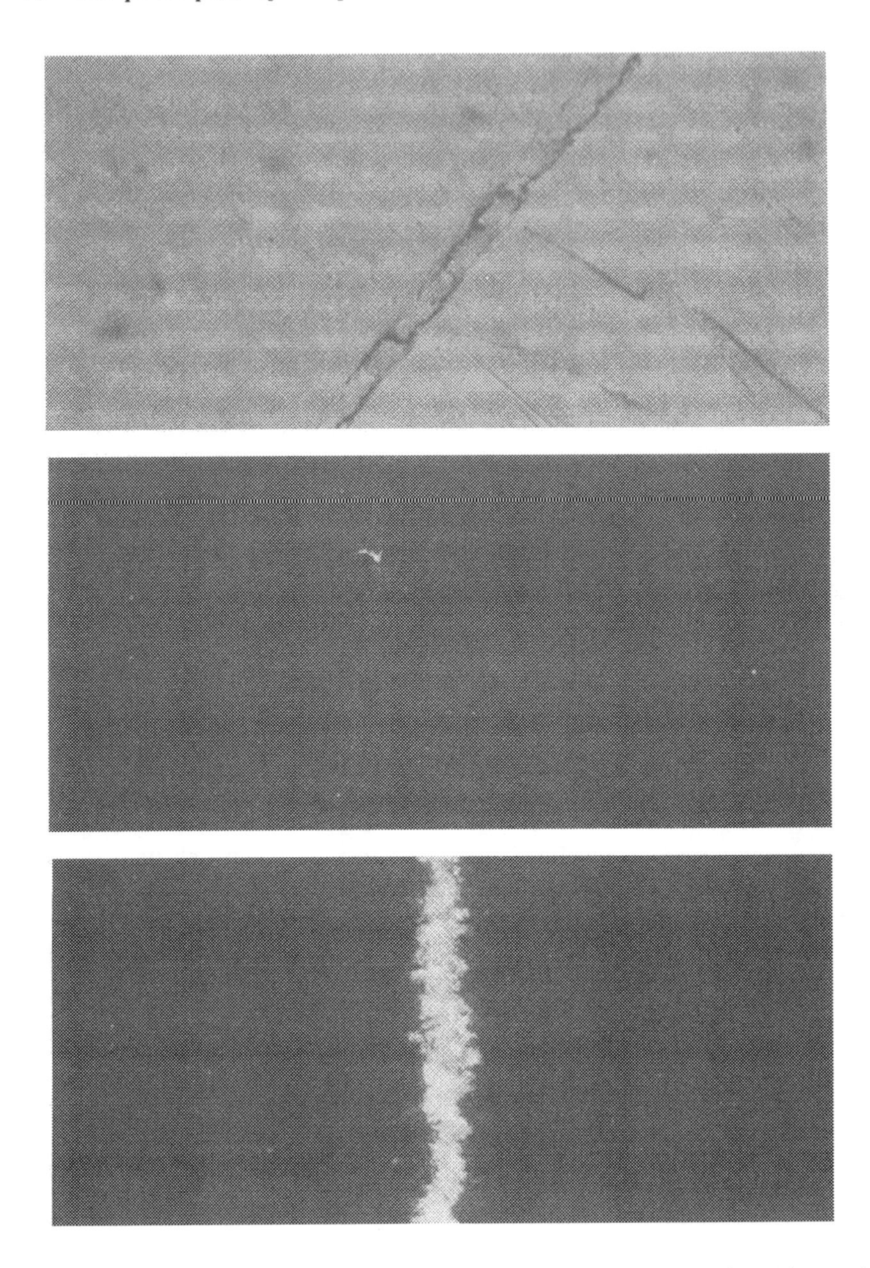

Fig. 4.3 The contribution of developers to the sensitivity of penetrant testing: (a) a crack in a test panel much magnified under white light which is quite obvious; (b) the same crack under the same magnification after processing with a high sensitivity fluorescent penetrant (since the test panel was photographed under UVA (black light) the crack is actually more difficult to see than in (a)); (c) the effect of applying a dry powder developer (the crack is now easily seen even without magnification). The dimensions of the crack are 1.3 μm wide and 50 μm deep.

which is 1.3 micron wide and cannot be seen with the naked eye. After processing with a penetrant system which uses a very bright and sensitive fluorescent penetrant the crack is still not seen by unaided vision.

After a developer has been applied and allowed a suitable time to work it is difficult to avoid seeing the indication of the crack.

All developers have two major functions:

1. to draw penetrant out of entrapment in defects and spread it across the surface
2. to enhance optically the visibility of the indication.

In addition to these functions the non-aqueous suspensions have a third function which is for the carrier solvent to invade the discontinuity, dissolve the penetrant residue and diffuse it back through the layer of developer pigments by capillary action where it is adsorbed on the pigment particles. In the case of colour contrast penetrant testing there is a fourth function which is to provide a contrasting background, usually white, against which the coloured indications can be seen. The penetrant provides the colour and the developer provides the contrast.

At this stage the phenomenon involved is capillarity followed by adsorption. However, the use of no developer, sometimes referred to as self-development, provides the initial action and an interesting insight into penetrants and their action. In the absence of a developer why should penetrant exude back to the surface? If the action whereby penetrants invades defects is capillary rise, the penetrant would not come back out after removal of surface excess unless a superior force was applied. A good developer might provide such a force but self-development of even large defects would not occur. Experiments with glass tubes and liquids will demonstrate this. When a liquid does not wet surfaces well considerable force is required to reverse capillary action, e.g. resetting a clinical thermometer. The theory of immersional free energy is much more attractive. When the surface excess penetrant has been removed the discontinuity contains an infinite concentration of penetrant and this is thermodynamically unacceptable. In an attempt to balance concentration of penetrant it floods back out to the surface to be seen or to be picked up by the capillary action of the developer particles.

Apart from having to explain away the fact that it should be easier to invade tight, fine defects than broad or medium ones, which is contrary to practical experience, the proponents of simple capillary action must explain self-development away. If capillary action took the penetrant in, what brings it out? Reverse capillary action is a convenient figment of the imagination – we cannot have it both ways. Forces need other forces to overcome them. The immersional free energy provides for such a force through entropy – capillary action does not.

4.3 AVAILABLE SYSTEMS

The available penetrant materials are listed in Table 4.1. Only the more commonly used materials are included and it is clear from simple arithmetic that, ignoring the various potential sensitivity levels available for the fluorescent penetrants or the various methods of applying the materials, there are 60 possible combinations to choose from. (This ignores the lacquer developers which are for special purposes only.) The problems of selecting a system are discussed in detail in Chapter 7. In this chapter only the outline schematics of the various systems are given.

Table 4.1 Available penetrant materials

Penetrant	*Remover*	*Developer*
Fluorescent	Water	Dry powder
Colour contrast	Hydrophilic remover	Non-aqueous suspension
Dual purpose	Solvent	Aqueous solution
	Lipophilic emulsifier	Aqueous suspension
		None
		(Lacquer)

4.3.1. Classification according to viewing conditions

Penetrants are formulated so that indications are viewed either in white light or in a darkened area under a UVA lamp. UVA is often referred to as black light which is a contradiction in terms but has been used for many years. Both terms refer to the region of the spectrum with wavelength between 315 and 400 nm. The peak output of these lamps falls between 360 and 370 nm which is ideal for exciting the fluorescent dyes in the penetrant. The dyes absorb the UVA and re-emit it at longer wavelengths. In the case of the standard fluorescent penetrants these fluoresce yellow-green and when excited emit that colour at a wavelength of around 520–550 nm. Dual-purpose penetrants appear pink or orange which is a longer wavelength.

Colour contrast penetrants normally give red indications. This is because red on white is the best chromatic contrast detectable by the human eye. The contrast of black on white is the easiest to detect. This is not used since many surfaces are themselves black and incomplete developer cover would increase confusion over false indications. Red does not suffer this disadvantage and offers good contrast in natural daylight or under artificial white light. Even colour-blind people will see red as a dark grey or black on white, although such people should not be carrying out inspection of penetrant-tested parts.

Dual-purpose penetrants can be used either as colour contrast penetrants whose indications can be viewed in white light, or as fluorescent penetrants whose indications can be viewed under UVA (black light) when they will fluoresce.

4.3.2 Classification according to method of penetrant removal

In this chapter we consider only the outline of the processes which are illustrated in the flowcharts in Figures 4.4 to 4.8. The pre-clean and post-clean stages mentioned in this flowchart have not been mentioned before in this chapter. Both are vitally important. Without proper pre-cleaning we cannot expect success with penetrant testing – any contaminant in a discontinuity will form

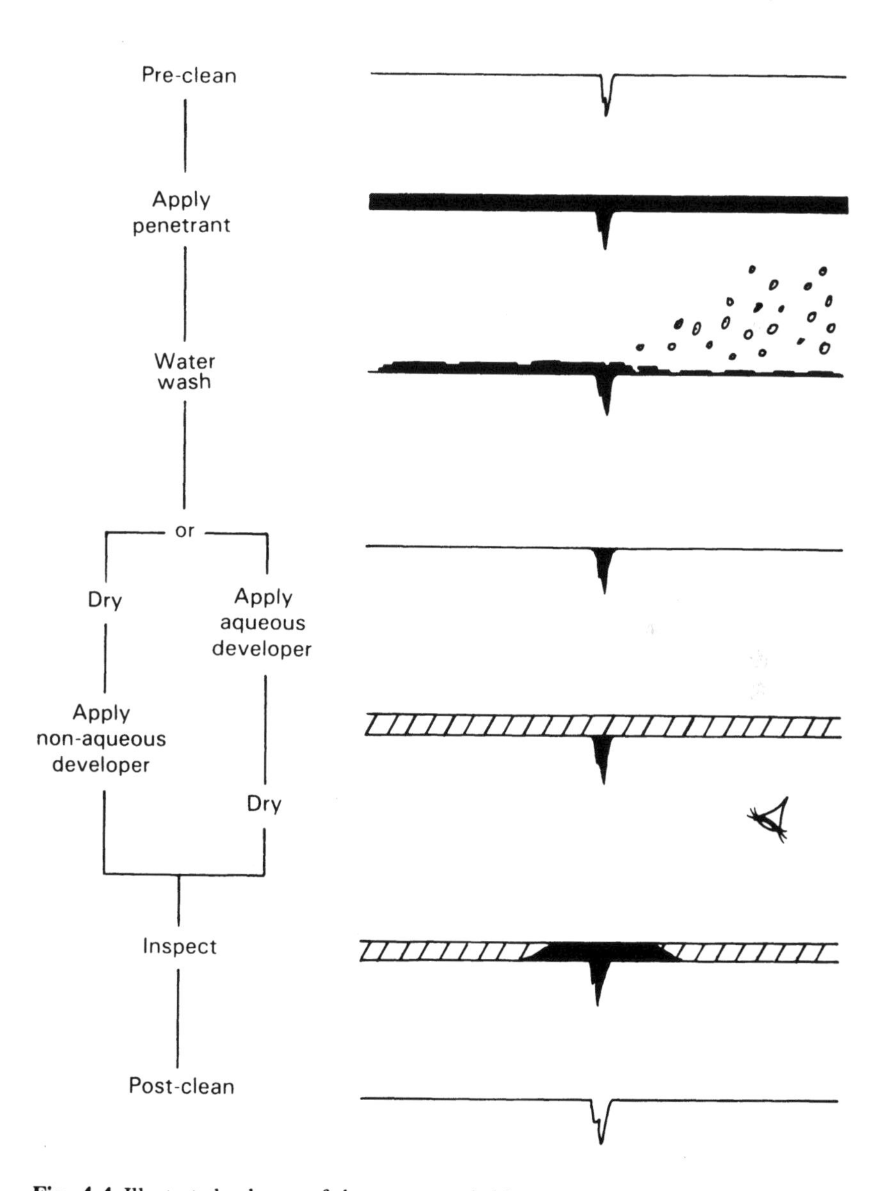

Fig. 4.4 Illustrated scheme of the water-washable process.

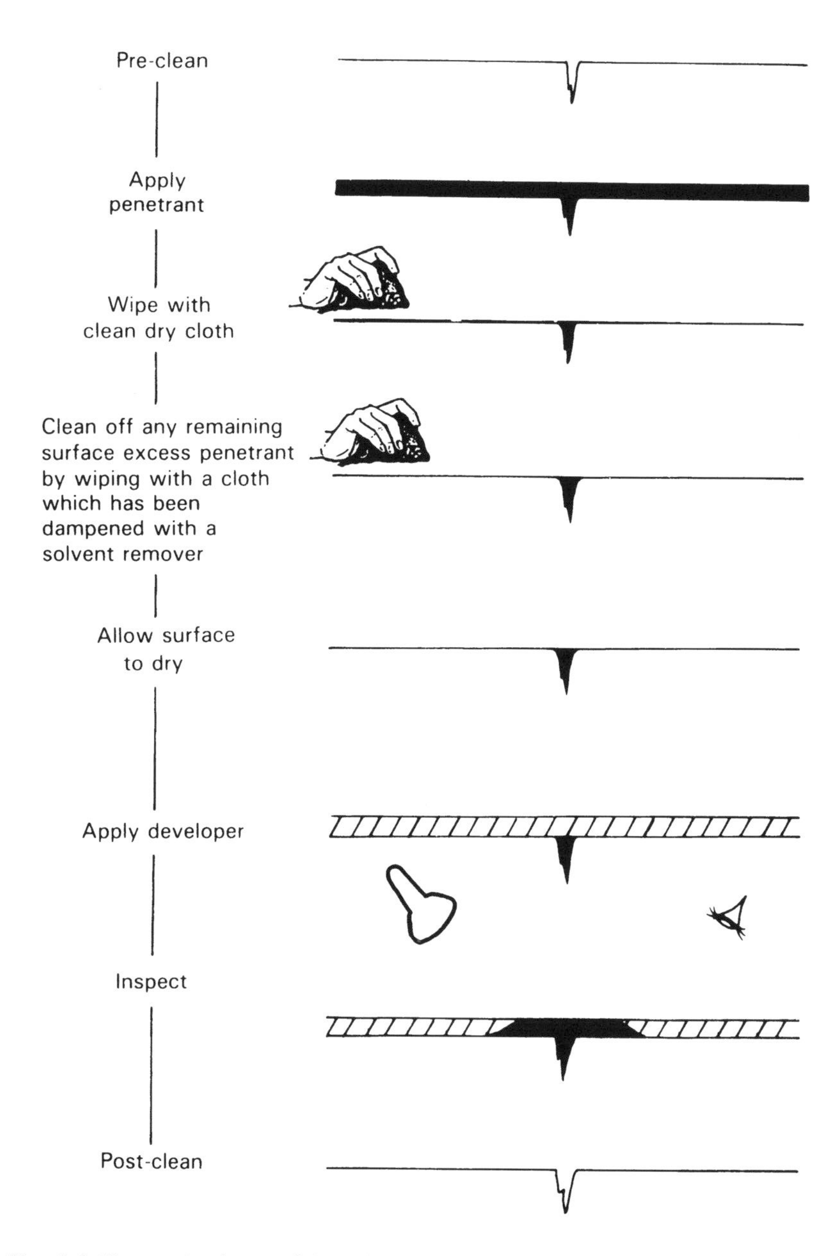

Fig. 4.6 Illustrated scheme of the solvent-removal penetrant process.

combination with immersion. Only in special circumstances is immersion alone satisfactory. Non-water-washable penetrant is commonly removed by one of the following methods:

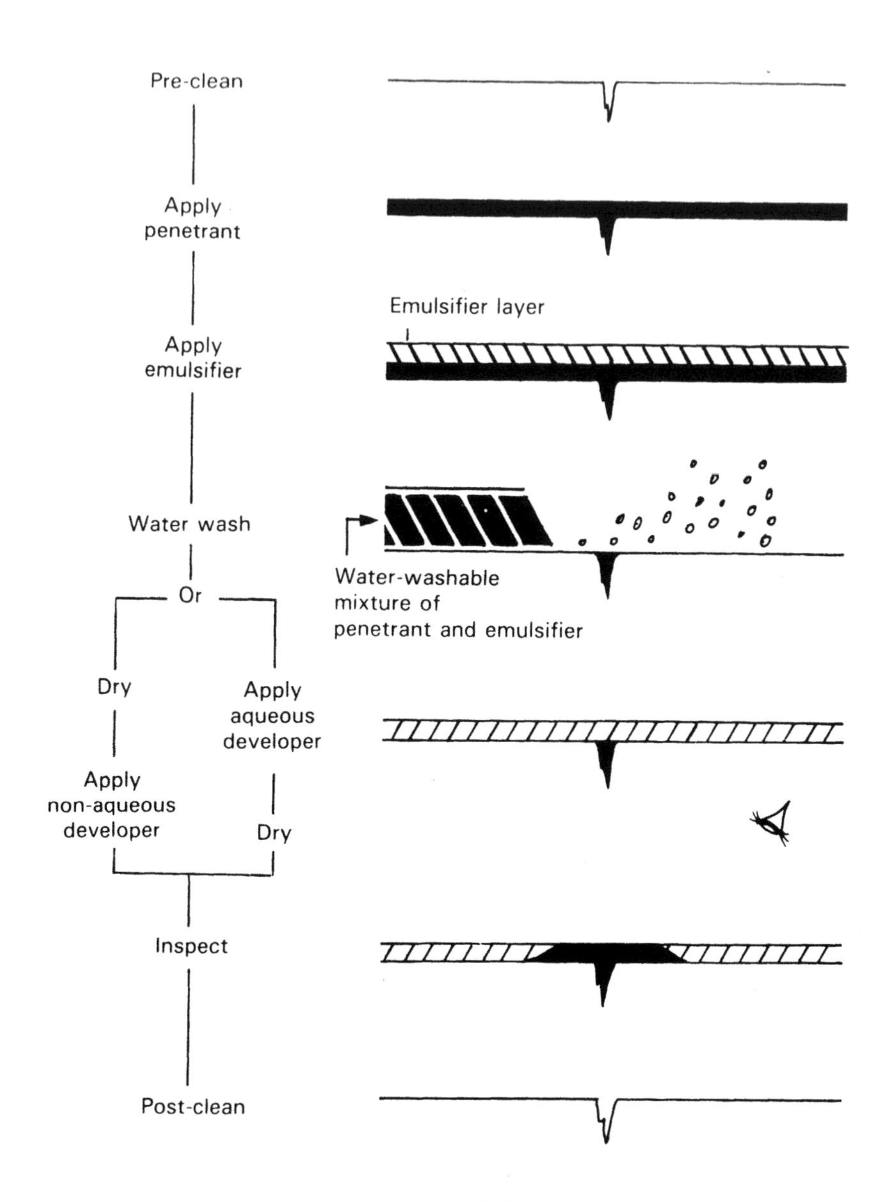

Fig. 4.7 Illustrated scheme of the post-emulsified penetrant process.

1. use of a hydrophilic remover solution (Fig. 4.5);
2. use of a solvent remover (Fig. 4.6);
3. use of a lipophilic emulsifer (Fig. 4.7).

The four commonly used methods for removal of surface excess penetrant are drawn together schematically in Figure 4.8.

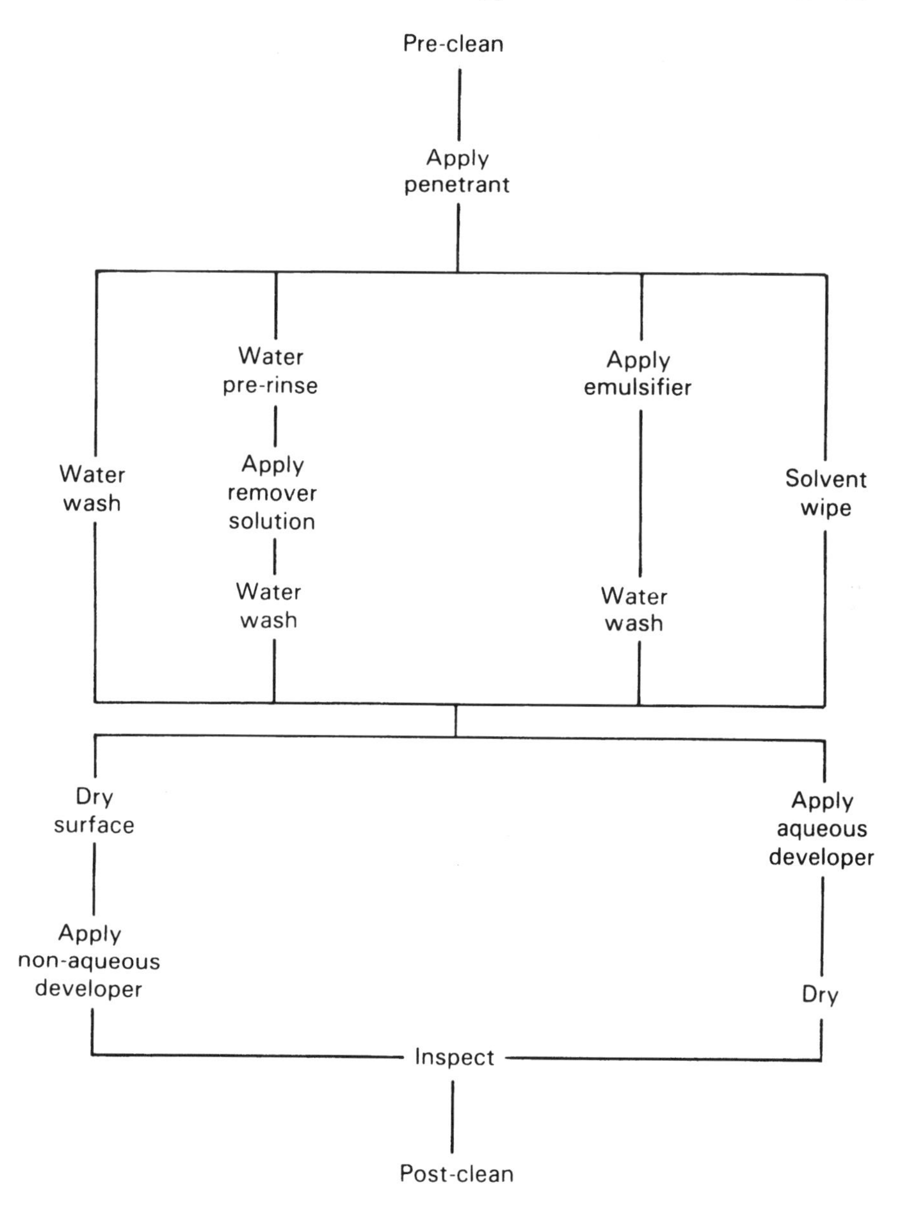

Fig. 4.8 General scheme of commonly used penetrant processes.

4.4 APPLICATIONS AND LIMITATIONS

Liquid penetrant testing can be applied to any essentially non-porous solid surface for the detection of discontinuities which are open to the surface. It is possible to detect surface cracks, porosity, through leaks, forging laps and bursts, shrinkage cracks, cold shuts, grinding cracks, heat treatment cracks, seams, defects at the surface of welds and lack of bond between two materials. Penetrant testing can be applied to metals including aluminium, magnesium, stainless steel, titanium, brass, copper, bronze, cast iron, steel, hafnium, tungsten, other metals and powdered metals.

Some examples of penetrant indications of various defects are shown in Figures 4.9–4.13. These examples include cracks in a coating, fatigue cracks and a casting defect which has remained undetected throughout its production. The materials in the four examples are a non-metallic coating, titanium and stainless steel. Even this small number of examples shows the versatility of penetrant testing.

Although penetrant testing is most frequently associated with inspection of metal surfaces it can be applied successfully to any non-porous solid surface. Examples of other materials which are tested using penetrant inspection include ceramics, glasses, carbide tools, plastics and composite materials. Engineering ceramics vary in density and, provided that they are sufficiently dense, penetrant inspection is an ideal method for searching for surface-breaking defects. When the surfaces of ceramic components are machined or fairly smooth the fluorescent penetrant post-removable hydrophilic process

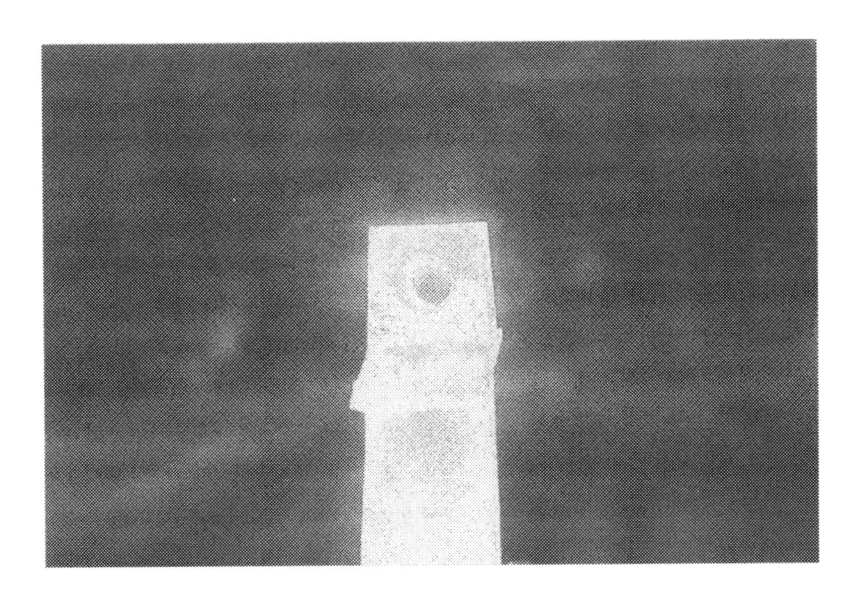

Fig 4.9 Indication of cracks in a ceramic coating on part of a turbine blade. Such components are readily recovered by stripping the coating and recoating. Failure to detect such defects would lead to problems during the service life of the component.

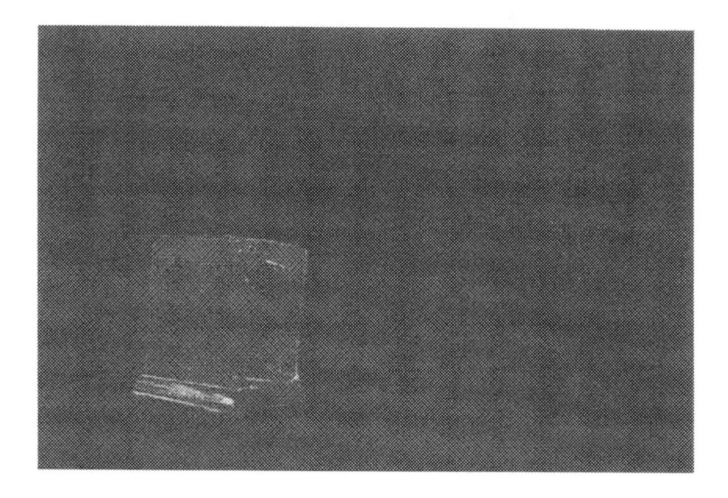

Fig. 4.10 Compressor blades have a hard life in service. This fluorescent indication shows evidence of serious fatigue in the material. If undetected, a piece of the blade will break away during service causing further damage.

gives excellent results. Interest has recently arisen in the use of penetrant testing of the ceramic heat shield coatings which are used increasingly on turbine components. The surface of such coatings is normally quite rough and special techniques, which are described in Chapter 5, have been found to be useful. Glass is a resistant material which is still very widely used. Application ranges from the humble bottle to advanced engineering components. Penetrant testing has been applied to bottles, particularly the tops, and to many other applications of glass including glass-to-metal seals. Carbide tools are often tested using the penetrant method, also with good results. Colour contrast processes or low to medium sensitivity fluorescent water-washable processes often give enough information for this application.

Penetrant processes are used on a wide range of plastic components and can also be used on composites for seeking surface-breaking discontinuities. Before any plastic or composite material is tested using the penetrant method, extensive compatibility tests must be carried out. Many penetrants contain chemicals which act as solvents for plastics and the resins which are used in many composite materials. Some penetrants contain

Fig. 4.11 Fatigue cracks at the bases of vanes in a section of a stator.

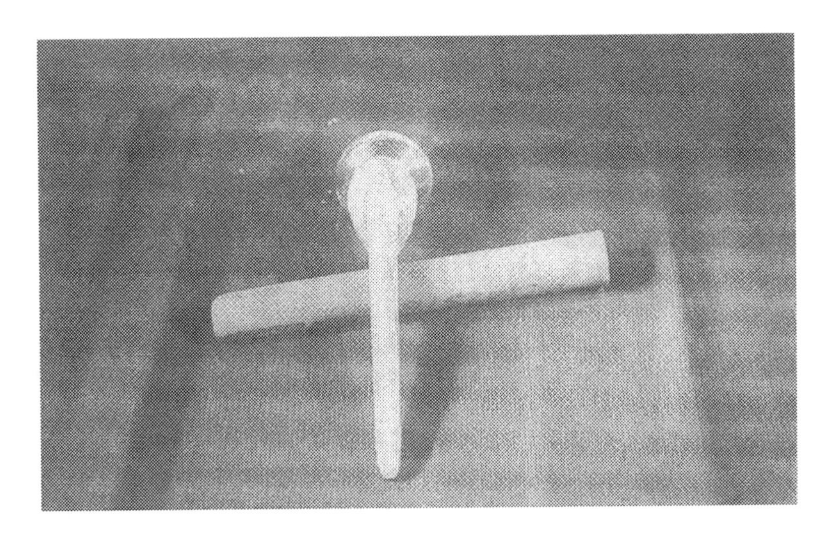

Fig. 4.12 This prosthetic hip joint would be subject to very significant stresses in use. An indication of a defect to the point shown here could lead to rapid failure of the component in use.

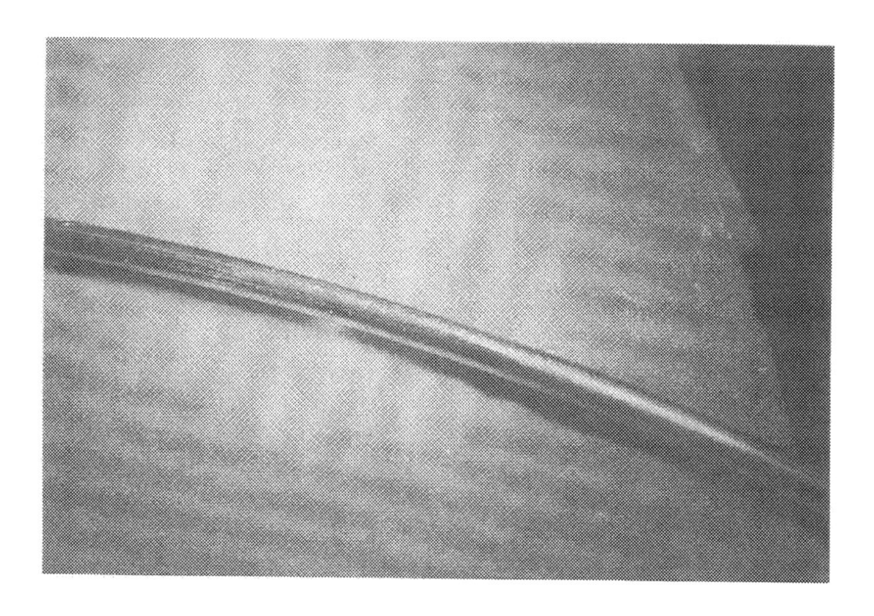

Fig. 4.13 Extrusion defects: fluorescent penetrant processing will indicate defects in extruded material as shown in this aluminium wire. However, the process of wire production is so rapid that penetrant inspection is not practical and an eddy current test is more appropriate.

chemicals which are used as plasticizers and these will have a disastrous effect on any vulnerable material. Immersion of scrap pieces of the plastic or composite material in candidate penetrants for 4 weeks with regular inspection for deterioration of the material is advisable. Penetrants which contain no hydrocarbons are available. However, this is not in itself a guarantee that the penetrant will not attack the material as other chemicals used in the formulation may be aggressive. A further point of caution is that the formulation of the plastic or resin may change, and even subtle changes of this type may make previously resistant materials vulnerable to chemical attack. Modern quality control procedures ensure that only very minor changes can be made to penetrant materials without a change in the product name, and so this aspect is not a threat.

The application of penetrant testing to plastic and composite materials must be carried out with continuous care over the point of chemical resistance of the material. Despite this complication, the penetrant method is used on these materials with great success and may be the only practical method for seeking defects which are open to the surface. The penetrant process is not strictly limited by the geometry of parts and fabrications, although some shapes do tax the ingenuity of penetrant engineers particularly when several surface states occur on the same part. The major limitation of the process is that it cannot be used for detecting internal defects nor will it work on a painted surface unless the paint is removed, and some other surface treatments, such as anodizing, can cause problems.

QUESTIONS

1. When carrying out a capillary rise experiment with tubes of different diameters in which tube will the liquid rise most?

   (a) They all rise to the same height
   (b) The tube with the largest diameter
   (c) The tube with the smallest diameter
   (d) The tube with an intermediate diameter

2. When considering application of penetrant testing of a component it is necessary to consider first

   (a) only the penetrant and its viscosity
   (b) the total process of penetrant, remover and developer
   (c) whether the penetrant is fluorescent or colour contrast
   (d) the choice of developer

3. Methods for removal of surface excess penetrant may involve

   (a) emulsification
   (b) detergency
   (c) solvent action
   (d) all of these

4. The most aggressive penetrant removal process is

   (a) water washing
   (b) use of a lipophilic emulsifying agent
   (c) use of a detergent (hydrophilic) remover
   (d) none of these

5. The two reasons for using a penetrant developer in *any* penetrant process are

   (a) to draw penetrant from entrapment to display it and enhance the visibility of the indication
   (b) to enhance the visibility of the indication and to suppress background

(c) to draw penetrant from entrapment to display it and to prevent rapid bleed out of the indication
(d) to ensure that indications of porosity become visible

6. The additional reason for using a suitable developer as part of the colour contrast processes is

(a) to suppress background
(b) to assist the post cleaning process
(c) to ensure a good background contrast for indications
(d) to leave evidence that the test was carried out

7. After processing surfaces with penetrant systems using fluorescent penetrant inspection is carried out

(a) in a dark area where white light is less than 20 lux (2 foot candles)
(b) with the aid of a UVA lamp (black light) giving an intensity of at least 1000 mW/cm^2.
(c) both (a) and (b)
(d) (b) is necessary but (a) is not

8. After penetrant removal by water washing or the use of either a detergent (hydrophilic) remover or an emulsifying agent (lipophilic) remover aqueous developers can be applied

(a) directly after the removal stage followed by drying
(b) only after the surface has been dried
(c) these developers can be applied to wet or dry surfaces as convenient
(d) both (a) and (c) are true
(e) each of (a)–(c) is true
(f) none of (a)–(c) is true

Note: There is only *one* correct answer to questions 1–8. Tick your choice and check it with the correct answers on p. 212.

9. Explain in a few sentences the difference between the action of emulsifying agents and detergent (hydrophilic) remover solutions.
10. When used correctly non-aqueous suspension developers contribute more to the sensitivity of penetrant processes than other types. Explain why this is and the fact that improper use of them can actually reduce the sensitivity of a process.
11. Explain why solvent removers must only be used in the wipe technique when they form part of a manual penetrant process.
12. Why is it important to keep strictly to manufacturers' and specification

instructions when making up solutions of detergent (hydrophlic) removers.

13. What are the two principal methods used for classifying penetrant processes.
14. Discuss the applications and limitations of the penetrant method of non-destructive testing.

5 Penetrant techniques

The techniques discussed in this chapter refer to typical liquid penetrant testing processes widely used in industry. Most of the sections apply equally to penetrant processes based on fluorescent penetrants, colour contrast penetrants and dual purpose penetrants; where there are differences specific to any particular process these are discussed in the appropriate section.

5.1 PREPARATION OF SURFACES

All methods of NDT benefit from thorough preparation. In the case of penetrant testing the proper preparation of the surfaces to be tested is so important that it is true to say that the success, or lack of success, of the whole process rests on how well surfaces have been prepared. Penetrant testing is applied either to raw material such as a billet of stainless steel, at an intermediate stage in manufacture, after new manufacture of a component or structure or after a component or structure has been in service. Generally it is true to say that parts brought to penetrant testing after service present the greater pre-cleaning problem. Parts which have been painted need paint stripped from them. Unpainted parts become dirty or corroded during use, and scale, accumulated carbon, oily dirt, rust or any other surface contaminant must be thoroughly removed. Also, anodizing and other surface treatments are best removed.

Most chemical methods of cleaning end with a water wash which is most important. Many cleaning and stripping chemicals are quite aggressive to fluorescent penetrants - chromates, acid ferric chloride, any concentration of nitric acid above 10% are extremely aggressive. Contamination of the surface by residues of such chemicals is likely to cause complete breakdown of the fluorescent penetrant processes. For this reason a thorough water rinse is essential after most chemical cleaning preparations.

After the water rinse it is equally essential that the surface is very thoroughly dried. Water forms an extremely effective barrier to the entry of penetrant into defects. No amount of water tolerance in the penetrant can compensate for this.

The golden rule of preparation of surfaces for penetrant testing is that they should be thoroughly clean and thoroughly dry.

5.1.1 Vapour degreasing

Until the mid 1990s vapour degreasing was the most popular method for the precleaning of component surfaces for penetrant testing. It offered, and still does, several practical advantages the most obvious of which is that the surfaces are dry at the end of the process as opposed to the use of any water based cleaners which leave wet surfaces. The most popular solvent used in these installations in recent years was 1,1,1, thrichloroethane which has a more acceptable health and safety profile than other volatile chlorinated hydrocarbons. It also has attractive properties as a solvent and it boils at a temperature which is very convenient for use in vapour degreasing. After it became firmly associated with the depletion of the earth's ozone layer its availability became first restricted and subsequently banned. A number of related compounds such as 1,1,1, trichloroethylene and perchloroethylene have similarly attractive practical characteristics and are not implicated in the damage of the earth's ozone layer however they do present somewhat sinister health and safety problems. These have been resolved to a great extent by developments in the design of vapour degreasing installations. The open top installations have been replaced with closed designs which have various levels of sophistication. These designs have allowed the acceptable application of vapour degreasing. Technically the process remains unchanged by the replacement of 1,1,1, trichloroethane with such materials as 1,1,1, trichloroethylene or perchloroethylene as they have similar solvency properties and boil at similar temperatures. The method removes only oily deposits from the surfaces of workpieces and leaves them dry at a temperature of around 80° C. It is normal good practice to allow them to cool to a temperature below 40° C before penetrant is applied. Apart from conforming to controlling specifications this allows any entrapped solvent to evaporate and also prevents penetrant drying out to any degree during the penetrant contact time. Some industries, including aerospace and nuclear engineering have always had concerns about the use of any halogenated materials on some metals, notably titanium alloys, nickel alloys, and stainless steels. These reservations are due to metallurgical concerns.

5.1.2 Solvent cleaning

Volatile organic solvents are widely used for cleaning surfaces in the manual testing of parts of structures on site, notably when aerosols are in use. Solvents are very useful for this. They are sprayed onto surfaces and then allowed to run off carrying oily contamination away from the area to be examined. Sometimes solvents are used in immersion tanks or even applied by spraying. A severe problem with this method is that the solvent becomes fouled with the contaminant and reaches the stage where as much unwanted oily dirt is deposited on the surface as is removed. To overcome this a distrillation process is often incorporated. A further problem is that volatile organic solvents which are useful for degreasing are flammable or harmful or both. If a solvent cleaning preparation is chosen, vapour degreasing has many advantages over solvent wash. With the loss of availability of 1,1,1, trichloroethane the only volatile solvents which have suitable characteristic are flammable since the harmful properties of the alternative non-flammable chlorinated solvents make them unacceptable for manual use. Solvents such as acetone, isopropanol, volatile hydrocarbons and methyl ethyl ketone should only be used in small quantities. Handling large quantities of such solvents safely is quite difficult. Fire hazards exist even at very low temperatures. Acetone has a flash point of around –10° C at these temperatures, the air is very dry and electrostatic discharges are common. Such sparks are more than sufficient to start impressive fires.

5.1.3 Detergent cleaning

Some form of detergent cleaning can be used to clean almost any surface or component. Detergents may be anionic, cationic, amphoteric or non-ionic. The most aggressive from the corrosion point of view are the anionic detergents and the least corrosive are the amphoteric and non-ionic types. These less corrosive types of detergent are just as effective as the anionic types and cost is no longer a serious problem. Pre-cleaners of all types should be non-corrosive to the surface on which they are acting. To this end inhibitors are formulated into the detergent cleaner. Detergency works best at temperatures around 80 °C and this can only be achieved in a washing machine. Cold or warm scrub, rinse or wipe techniques can be used to remove light contamination. Above 40 °C the mechanism of detergency starts to become more aggressive and the introduction of heat to the system aids the cleaning process. After detergent cleaning surfaces must be rinsed with clean water and dried. Drying must be both hot enough and long enough to ensure that all water is driven from the surface and the defect.

5.1.4 Steam cleaning

Most steam cleaning chemicals are quite strongly alkaline although some are more neutral. For large parts or structures steam cleaning is ideal. An application which is very successful is on aircraft landing gear. Unfortunately some of the early chemicals sold for steam cleaning were quite aggressive and the process gained a poor reputation. As with any process involving water, the surface must be rinsed with clean water and then dried thoroughly before penetrant testing is started.

5.1.5 Paint removal

Paint is a very important protective for a large number of structures and components. Paints have become very sophisticated indeed and are now specifically prepared for almost individual uses. The resistance to both physical and chemical attack can be very great. As paints have developed, so too have removers. Paint removers may be formulated to be used as liquids in hot or cold tanks; they may also be made up as a thixotropic gel to be smeared over the painted surface. The latter type is termed an 'application paint stripper'. These types of stripper may be acidic, neutral or alkaline.

Any type of paint removal which does not harm the test surface is acceptable for preparation for penetrant testing. Physical means such as grit blasting and wire brushes are best avoided as they often interfere with the nature of the solid surface. Chemical paint removal is always to be preferred. Few paints are readily removed by a simple solvent, leaving a clean dry surface. Most paint strippers together with any residual removed paint are washed off with a water rinse. Before penetrant testing these surfaces must, of course, be dried.

5.1.6 Rust and scale removal

Rust removal and scale removal are often considered together. Rust is relatively simple to remove chemically with alkaline or acid solutions. Care must be taken in the choice of these chemicals as they can cause problems on metallic surfaces. Rust is found only on iron and steel, and so alkaline solutions are normally quite acceptable. Acids can be used but only with care. Surface scaling can be very difficult to remove and frequently needs a multi-stage process incorporating acid treatments, alkaline processes and oxidizing steps, all with water washing between them. Great care must be taken to ensure that no descaling solution is carried over into the fluorescent penetrant as this will reduce fluorescent brilliance significantly – in some cases it will destroy it completely.

Physical methods for rust and scale removal prior to penetrant testing are to be avoided. Such methods, particulary if effective on scale, can be expected to smear metal so that open defects are closed or to fill them with particles of contaminant, grit or wire brush.

5.1.7 Ultrasonic cleaning

At one time ultrasonic cleaning was held in very high regard as method for preparation of parts for penetrant testing. It is often used in conjunction with a solvent or a detergent bath to improve efficiency. Both solvents and detergents are very good at removing organic surface contamination. Ultrasonic cleaning is very useful when hard solid matter can be shaken loose and either dissolved in the surrounding liquid or floated off. However, if residues on the surface are fairly plastic – as is the case with many modern paints – the ultrasonic shock will be absorbed and the contribution to the cleaning process will be minimal.

Ultrasonic cleaning can be very useful but is not the ultimate answer that it was once thought to be.

5.1.8 Etching

When etching is permissible it is probably the best method available for preparing surfaces for penetrant testing. Etching solutions open up the edges of surface discontinuities, thus allowing easier access for penetrant at a later stage. The process of etching has been used as method for testing for surface-opening defects. In an etch solution the edges of a discontinuity are attacked preferentially, and after a rinse these areas are more easily seen by the naked eye or under low power magnification.

Clearly, if any loss of material is unacceptable or if the etch solution is likely to raise problems of hydrogen embrittlement in the material the process is not acceptable. The water rinse following an etch must be very thorough as etchant chemicals are often very destructive of fluorescent brilliance.

5.1.9 Mechanical methods

Mechanical methods are sometimes termed physical methods, but since solvency is a physical phenomenon the term mechanical is preferred here.

A basic rule of the use of mechanical cleaning methods is that when they appear to be the only way of cleaning surfaces think again – when this has been done three times and the conclusion to use a mechanical method cannot be avoided, start to think of ways of overcoming the problems that mechanical methods cause. Mechanical methods often involve an abrasive process which uses material which is harder than the base metal. This leads to smearing of the metal (Fig. 5.1) which will reduce the opening of a defect or even peen it over and close it completely. A secondary problem can be that the abrasive medium may itself become trapped within defects, either alone or together with some of the contaminant from the surface.

Fruitstone grits and ground nutshells have been used with great success both alone and in combination with wet chemicals for cleaning purposes. These

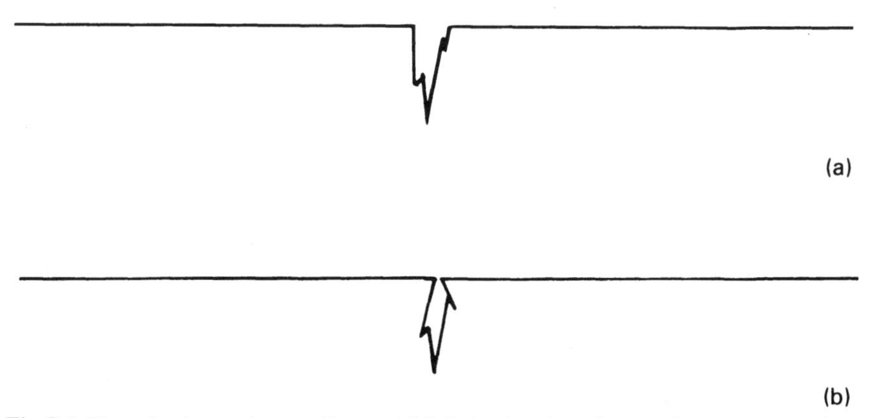

Fig 5.1 Use of grit or other medium, which is harder than the metal, causes smearing which can reduce or even close the opening of a defect: (a) a defect open to the surface; (b) after cleaning the surface with a medium which is harder than the metal the defect can become closed totally by smearing of the metal.

materials are, of course, softer than metals. The major danger with these products comes from the very real danger of using the wrong grade of grit (this *has* happened). When a mechanical method of surface preparation must be used it is advantageous if it can be followed by an etch process to overcome the problem of closing up defects.

5.1.10 Drying surfaces

The need for surfaces to be dried thoroughly before penetrant is applied to them cannot be over emphasised. Water is as much a contaminant in this context as anything else and is more commonly found. It is not enough that the surface should be visibly dry as this may be so when water remains entrapped in discontinuities. Experimentally it has been found that when surfaces are dried at a temperature of around 120° C for 60 minutes they are satisfactorily dry for the penetrant process. This method of drying surfaces has a number of disadvantages. First is the obvious delay in the process with 60 minutes replacing the around 10 which was previously needed, this takes no account of the extra time which is needed, this takes no account of the extra time which is needed to cool the parts from 120° C to 40° C as opposed to cooling from 80° C. The actual cooling time will vary according to the mass of the workpieces but is unlikely to involve less than 30 minutes extra. It is of no use to apply cold air to the surfaces as this will cause condensation from the atmosphere to become deposited on the surfaces which will them become wet again. A further problem is that such treatment of heating some alloys for that length of time will cause metallurgical changes which are unacceptable. A practical solution is for the workpieces to be dried under reduced atmospheric pressure in a vacuum oven. The speed of drying under reduced pressure can be calculated theoretically however there are pitfalls to be avoided. The levels of vacuum which is necessary to allow practically advantageous drying times are such that there is a strong possibility of water freezing on the surfaces. This can be overcome if some heating of the workpieces is available. The heat can be applied either by immersion of the workpieces in warm water before passing to the vacuum oven, provision of radiant heat in the vacuum oven, or a combination of both measures. When such precautions are taken it becomes possible to maintain a practical drying time of 10 minutes in a vacuum of 10 millibar if the temperature of the workpiece surface is maintained at 40° C. Vacuum ovens need to be of robust construction in order to withstand the external atmospheric pressure each time they are operated. They must also be leak proof and require substantial pumping apparatus. They represent a substantial initial cost however their robust construction and powerful pumps tend to give long life with few maintenance problems. This offsets the initial investment to a significant degree.

5.2 APPLICATION OF PENETRANTS

It really does not matter how penetrant is applied to part surfaces provided that it covers the whole area under test. Methods used include immersion, flow on, brush on or wipe on, aerosol spray, conventional spray, electrostatic spray or application as a fog.

5.2.1 Immersion

Immersion of parts in penetrant has the great advantage that it ensures that all surfaces of a component are covered. In the case of parts which have a complicated shape immersion can cause problems. Large quantities of penetrant may become trapped inside hollow parts which must therefore be turned to allow the material to drain out before moving to the next stage. Not only will such retention of large quantities of penetrant lead to poor economy but it will also interfere with the process for removal of surface excess penetrant. Blind holes or even through holes can be difficult to drain after immersion, and this can also interfere with the rest of the process. Small hollow parts with complex internal forms, e.g. turbine blades with cooling holes in them, can present insoluble difficulties if penetrant is applied by immersion. In such a case penetrant enters the small cooling holes in relatively large quantities but does not drain out. Some of it is removed in the removal process, but so much remains that it bleeds back out during drying and development making inspection of the areas around the cooling holes impossible because of excessive background. The most successful solution to this problem is to use an electrostatic spray (section 5.2.5). When the immersion technique is used to apply penetrants it is best to allow at least 3 min (some authorities require 5 min) actually in the penetrant before withdrawing the part and allowing it to drain for the remainder of the contact time. Complicated shapes and blind holes can create other problems when penetrant is applied by immersion; they may form air traps so that no contact at all is gained with the penetrant. For this reason it is useful for components to be moved about in the penetrant bath to break such air pockets up. Even when this is done it is a wise precaution to check areas where air pockets may occur during penetrant drainage to ensure total cover of the surfaces. The subject of penetrant contact time is discussed in section 5.2.6.

5.2.2 Flow on, wipe on and paint on

Flow on, wipe on and paint on are considered together because they are all useful for localized inspection or inspection of small numbers of parts which can be dealt with manually. Of the three, flow on has the attraction that it avoids the use of a brush or absorbant pad which could later become contaminated with another penetrant or even another chemical. Once penetrant is in the hairs of a brush or the fabric of a swab there is no way of removing it. Absorbant pads can be fitted in the caps of small tins of penetrant, thus reducing the possibility of cross contamination. Separate brushes or swabs can be expected to be used for a variety of materials with unpredictable results.

5.2.3 Aerosol spray

Aerosol spray is almost certainly the most common way in which colour contrast penetrants are applied. However, fluorescent penetrants are relatively rarely applied in this way. The main reason for this is that critical parts and portable parts are nowadays normally inspected by use of a fluorescent penetrant system, thus taking advantage of the higher sensitivity available from these methods. Colour contrast penetrants are very useful where the darkness needed for inspection of fluorescent penetrant indications is difficult to obtain. This means that almost all inspection of fixed structures on site at power stations on oil rigs and in similar situations is carried out using colour contrast penetrant and most of it is applied by aerosol. Aerosol spray is technically quite an acceptable method of applying penetrant. It has the advantage of being readily portable for site work. When working outdoors with aerosols it is advisable to check the direction of the wind first as a faceful of sprayed penetrant is quite unpleasant.

5.2.4 Conventional spray and fog application

Large parts with simple shapes and medium sized parts with quite complex shapes are often suitable subjects for application of penetrant by conventional spray techniques. Conventional spray is rarely used since it tends to deliver too much material compared with electrostatic spray which offers a number of possibilities. Another method of applying penetrant by way of airborne droplets which are not electrically charged is to create a fog. The very fine droplets of a fog wet surfaces very well. However, this method can only be used in a closed cabinet as it cannot be controlled directionally . At present there is too little reported experience on fog application of penetrant for an objective assessment of its advantages and limitations to be made.

5.2.5 Electrostatic spray

Harold Ransburg investigated electrostatic spraying of liquids as long ago 1937. During the Second World War the process was developed steadily for application of paint and other protective coatings. There have been a number of phases in the electrostatic spray application of penetrants over the last 20 years. At one time it was believed that the method was universally applicable. However, after the initial discovery that there could be difficulties enthusiasm waned. This stage was followed by a more cautious approach which brings us to the present stage where electrostatic spray application has very significant advantages for specific purposes.

The theory of electrostatic spray is that the material – in our case penetrant – is reduced to a very fine spray which is then given a strong electrostatic charge. The applied charge ranges from 40 to 100 kv. This charge is imparted to the droplets which are then expelled from the electrostatic gun and travel to the earthed target where they land, lose what charge is left and coalesce to form a layer. In practice the phenomenon is more complex and the conductivity of the liquid plays a significant part in the process.

Three basic types of electrostatic gun are available:

1. air-assisted guns
2. totally electrostatic guns
3. pressure-assisted guns

Pressure-assisted guns do not interest us with regard to penetrant testing. They are suitable for coating enormous areas such as the sides of ships. The remaining two types of gun are of interest to penetrant testing. Some guns use totally electrostatic separation where the liquid flows across a surface to a narrow edge. The surface is highly charged and imparts a like charge to the liquid so that when it reaches the edge it is electrostatically repelled in the form of tiny droplets. This type of gun is very useful provided that the liquid has a suitable specific conductivity within a fairly narrow range. Air-assisted guns use air to atomize the liquid and impart an electrostatic charge as the liquid leaves the gun. Conventional electrostatic guns are available which spray liquids of a wide range of specific conductances. Recently guns have become available which use an air turbine to generate the electrostatic charge, and these also work with liquids across a wide range of specific conductances.

When electrostatic guns are used to spray penetrants which display a very wide range of specific conductances it is very desirable to use guns which can deal with a wide range of materials.

Specific situations where electrostatic spray application has definite advantages include the situation where large parts with simple shapes such as the wing stringers of aircraft need to be inspected by penetrant testing. The large area can be covered with penetrant quickly, efficiently and economically. The second situation has also been mentioned in the discussion of immersion in section 5.2.1. Electrostatic spray is a very useful technique when the parts to be inspected have a number of small blind or through holes where penetrant can be trapped and bleed back causing heavy background at inspection. An extreme case is that of small hollow turbine blades with cooling holes in the leading and trailing edges. An appropriate technique using electrostatic spray application of the penetrant can allow the eventual developed part to be inspected very easily. This phenomenon is due largely to the Faraday cage effect. When a large number of like charged particles arrive at a surface near a hole they seek the nearest and strongest earth which is the corner of the hole. Also, when material does penetrate into the hole it repels other particles of like charge, thus limiting the quantity of penetrant entering the holes. Defects are far too small for a Faraday cage effect to operate. It has been claimed that electrostatic application leads to increased sensitivity of the penetrant because of the electrical nature of the process. These claims are difficult to substantiate and probably reflect the fact that the process is performed with increased care.

A practical point which must be considered is that when penetrant is applied by electrostatic spray a very thin layer is quite sufficient and parts should be checked with a UVA lamp (black light) after penetrant application to ensure that full cover has been achieved. If the surface drips penetrant, too much has been applied and much of the advantage of the process is lost.

Table 5.1 Recommended penetrant contact times

Material	*Form of of component*	*Contact time (min)*						
		Cracks	*Fatigue cracks*	*Shrinkage cracks*	*Porosity*	*Cold shuts*	*Laps*	*Lack of bond*
Aluminium and its alloys	Cast	10	20	10	10	10	–	–
	Forged	10	20	–	–	–	10	–
	Welded	15	20	–	15	–	–	15
Magnesium and its alloys	Cast	10	20	10	10	10	–	–
	Forged	10	20	–	–	–	10	–
	Welded	15	20	–	15	–	–	15
Stainless steels/ high nickel or cobalt alloys	Cast	15	30	15	15	15	–	–
	Forged	20	30	–	–	–	20	–
	Welded	20	30	–	20	–	–	20
	Tools	20	30	–	20	–	20	–
Titanium and its alloys	Cast	20	30	20	20	20	–	–
	Forged	25	30	–	–	–	25	–
	Welded	30	30	–	30	–	–	30
Brasses and bronzes	Cast	10	20	10	10	10	–	–
	Forged	10	20	–	–	–	10	–
	Brazed	15	20	–	15	–	–	15
Tungsten	Cast	60	90	60	60	60	–	–
	Extruded	60	90	–	–	–	–	–
Ceramics	Cast	20	30	–	15	–	–	–
	Machined	15	30	–	–	–	–	–
	Coatings	30	–	–	–	–	–	–
	Tools	20	30	–	15	–	–	20
Glass	All forms	10	–	–	–	–	–	–
	Glass/other material seals	10	–	–	–	–	–	20
Plastics	All forms	10	30	–	–	10	–	20
Metallic coatings	All forms	30	–	–	–	–	–	–

These times are suggested in the absence of clear instructions in a procedure or specification.

5.2.6 Penetrant contact time

In Chapter 4 it was indicated that penetrant activity is a physical phenomenon. Therefore penetrants need time to work. The amount of time needed to achieve the best results from a penetrant depends largely on the type of material being tested, the type of defects sought and the state of the surface. Temperature affects all physical processes. Fortunately, penetrants normally work within acceptable contact times at temperatures between 10 °C and 40 °C. At temperatures below 0 °C problems arise and special materials are needed. Between 10 and 40 °C penetrants can be expected to invade most defects in most materials within an hour. The difference between contact times is quite remarkable. Penetrants will invade defects in aluminium- or magnesium-based metals very quickly, whereas stainless steel, titanium, hafnium and tungsten present a different problem and longer contact times are needed. As a general rule contact times of less than 10 min are best avoided as being too short. Table 5.1 gives a list of minimum recommended penetrant contact times for various materials and defects. It is usually best to establish contact time for a specific problem experimentally wherever possible.

5.3 WATER WASH PROCEDURES

In common penetrant processes water washing is used in four ways:

1. in the removal process of a water-washable penetrant;
2. as a pre-rinse in the hydrophilic removal penetrant process;

5.3 Spray washing

Spray washing can be performed by using either a water spray or an air–water spray. In each case care must be taken to control the spray pattern and water droplet size and velocity. Spray patterns may be in the form of a fan or a cone, but strongly directional jets should not be used. Very fine sprays which tend to form mists are difficult to control and very heavy water droplets can cause irregular washes. A spray of moderately heavy droplets at moderate speed gives satisfactory results.

Water temperature should be controlled, ideally around 25 ± 5 °C although variation in the range 10–30 °C can be accepted. Pressure should not be greater than 2.8 bar (40 p.s.i.) and spray nozzles should be kept 30 cm (12 in) from the surface, particularly on manual lines. No part of the surface should be sprayed for more than 2 min.

5.3.2 Immersion wash

Immersion of parts covered with penetrant has one very severe shortcoming – the water becomes severely contaminated very quickly. Even with an efficient weir system this problem is not overcome at all in the case of water-washable penetrants and the wash station rapidly becomes a foaming mass of fluorescent green or dark red liquid of dubious and uncontrolled properties. In the case of water-washable penetrant systems or pre-rinses immersion is best considered as an adjunct to spray rinsing. In the case of washing after application of a hydrophilic remover solution or an emulsifier the immersion wash is used to stop the removal process, particularly when large parts are being processed and if a hand-held spray rinse were used some parts of the surface would have far longer contact time than others with the penetrant remover.

There are special situations when complex shapes are to be inspected where water-washable penetrant can be very successfully removed by immersion in agitated warm water (30–35 °C). At this temperature the water forms a clear solution of penetrant and will absorb practicable amounts. There are natural drag-out losses which are compensated with clean water, thus keeping the concentration of penetrant contamination down.

Before moving to the specific cases where water wash is used one comment is necessary. At all water rinse stages in a fluorescent penetrant process it is desirable to have UVA (black light) illumination to assess the success of the operation. At the stages of washing a water-washable penetrant, rinsing after application of a hydrophilic remover solution or rinsing after application of a lipophilic emulsifier it is *essential.*

5.3.3 Removal of a water-washable penetrant

The removal of surface excess penetrant is the most vulnerable stage of

Fig 5.2 An example of the results of poor penetrant removal. The 'tide marks' on the aerofoil surfaces of this segment are unlikely to be mistaken for defects, except in the case of the right-hand one. Poor removal is most likely to occur when water-washable penetrant processes are used as in this case. It always causes difficulties at inspection and must be avoided.

all penetrant processing in every system. It is a particular problem for water-washable penetrant systems (Fig. 5.2). In automatic penetrant lines the wash stage conditions can be fixed experimentally before the system is put to production use. In manual lines this stage is open to wide abuse. Actual technique varies according to the size of parts and the way in which they are presented at the wash station. Large parts often arrive on a hoist. Spray rinsing as outlined in section 5.3.1 should be used, i.e. spray head no closer than 30 cm (12 in), water temperature 25 ± 5 °C, pressure no greater than 2.8 bar (40 p.s.i.) and no part of the surface to be sprayed for more than 2 mins. Spray washing the bottom first and working upwards is a commonplace and satisfactory technique. Smaller parts often arrive singly or in small numbers in baskets, and these must be turned to ensure a complete wash. Small parts such as bolts or fasteners arrive in baskets, usually in very large numbers. Production managers naturally want as many parts through a line as quickly as possible. Where possible NDT departments must, of course, co-operate; however, this is an instance where the production manager should be overruled. If baskets are overloaded with small parts it is not possible for the operator to wash surface excess penetrant off evenly. Often a basket will be overfull, and the outer parts will be washed but on inspection the parts situated on the inside will be covered with residual background. If the inspector does as he should and returns the processed parts for reprocess it is very probable that there will be no background the second time and there is a fair danger that the wash will have been so thorough that some indications will also have been washed out. More than two or three layers of small parts in a basket on a penetrant line is totally unacceptable. The operator must be allowed to wash parts efficiently and check the level of wash under a UVA lamp (black light). A small degree of background is more desirable than a totally clear surface. Even indications of small defects can be seen through a light background, and it is an indication that the part has not been overwashed. So far in this section no mention has been made of immersion rinsing. This is because, when used alone, it is rarely satisfactory. A strong case can be made for making spray washing mandatory for castings. Immersion rinses can be useful as an assistance to spray rinsing, but except in special circumstances they are not very useful for removal of water-washable penetrant. The water becomes contaminated during the first use and this introduces an uncontrolled and uncontrollable element into process. Results are often unsatisfactory and problems of streaky finish and incomplete removal lead to severe problems during inspection.

5.3.4 Water pre-rinsing in the hydrophilic removal process

In water pre-rinsing much of the penetrant is driven off physically. Provided that the penetrant is truly non-water-washable there is no emulsification. This spray rinse leaves a thin uniform layer of penetrant to be removed. Well over 90% of the surface excess penetrant is removed by an effective water pre-rinse which is fortunate as the solutions of hydrophilic remover have very low tolerance for penetrant (less than 5% by volume at the concentrations used) and would become exhausted very rapidly without this step. A secondary advantage which is becoming more important is that the effluent from the pre-rinse can be separated readily and the water recycled.

The conditions for water pre-rinsing are much the same as those for spray rinsing of water-washable penetrant with regard to temperature, pressure and distance between the gun and the part. Time is much less critical in this case. Immersion pre-rinses alone are unsatisfactory in this instance, just as they are for removal of water-washable penetrants but for different reasons. First, they are inefficient even with agitation; immersion rinses leave around 50% of the original surface excess on the parts. Second, although the penetrants used in this process do not mix with water chemically, even with a good weir system the water rapidly becomes fouled with penetrant and the process becomes even less efficient. Immersion can be very helpful as an adjunct to spray rinsing when the parts being processed have complicated shapes.

5.3.5 Post-rinsing after application of hydrophilic remover solution

Application of any active penetrant remover must be controlled carefully if satisfactory results are to be achieved and repeated. In order to maintain control of the process contact with hydrophilic remover solutions is restricted to 2 or 4 min depending on the type of penetrant used. The quickest and most practical way of stopping the activity of penetrant remover is to dilute it with an infinite volume of water. When small parts are being processed this can be done by spray rinsing. If parts are large or there is a large number of them it is essential that they are immersed in a tank of cold water to stop the remover action. The volume of water must be sufficient to avoid the build-up of too great a concentration of remover over a working period and the creation of a second remover bath. When a penetrant line is in constant use the water should be changed after 24 hours of working. A spray rinse using standard conditions should be used to wash off surfaces after immersion.

5.3.6 Water rinse after application of a lipophilic emulsifier

The need for water immersion to stop the action of lipophilic emulsifiers is even greater than in the case of hydrophilic removers. Lipophilic removers are very active in emulsifying penetrant and can invade defects very rapidly and emulsify the penetrant trapped in them. Some lipophilic removers act very rapidly and emulsifier contact times can be as low as 45 s. In such cases an instant kill is very important if the sensitivity of the system is to be retained. In the case of lipophilic emulsifiers the water becomes contaminated, as it does in the case of hydrophilic removers, and should be handled on the same basis. Spray rinsing under standard conditions should be used after immersion.

5.4. APPLICATION OF HYDROPHILIC REMOVER SOLUTIONS

Hydrophilic remover solutions are used at concentrations ranging from 5% to 30% according to specification, application is normally by immersion and specifications vary as to whether or not the solution can be agitated. Contact times in hydrophilic remover solutions may be 2 or 4 min depending on the penetrant in use. Very bright penetrants require longer immersion time. In recent years much more interest has been shown in application by spray. This calls for much lower concentrations down to 0.1% and up to 2.5% depending on the brightness of the penetrant and the state of the surface. Spray application offers the advantage that the pre-rinse step can be omitted since the sprayed-on solution is washed away and not re-used. There are some installations which employ electrostatic spraying of hydrophilic remover solution. Theoretically it is possible to adjust the removal characteristics of detergents by changing the electrostatic charge on the gun. At present too little practical experience exists to quantify this phenomenon.

Foam application of hydrophilic removers has been used for many years. This is particularly attractive when the inside of a bore needs to be inspected. The classic example is an impeller from a jet engine. The inside bore is smooth and the area is critical, calling for a post-removable penetrant. Controlled removal of surface excess penetrant by conventional means is difficult to ensure. If the remover solution is pushed through a sinter using air it will foam. If such a device is placed under the part (Fig. 5.3) and foam is bubbled up through the bore good even removal is achieved. Foam application can also be achieved using an aerosol spray, and this method gives excellent sensitivity for critical inspection of parts where a manual technique is needed.

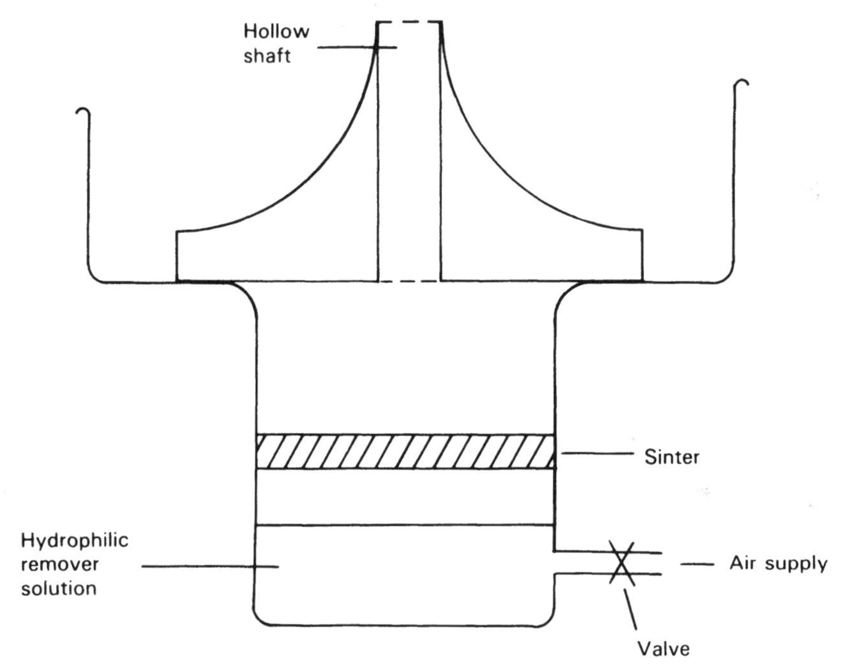

Fig 5.3 Foam application of hydrophilic remover solution: the air is bubbled through the remover solution and creates foam which passes through the sinter into the tank above. The component shown has a hollow shaft (typical of a turbine compressor) through which the foam can pass, alternatively the components can be held above the sinter and foam passed over them.

5.5 NORMAL PROCEDURES FOR USING SOLVENT REMOVERS

Solvent removers are the most aggressive type of all as far as the penetrant is concerned. Consequently the greatest care must be taken in their use. With proper care and use very good results can be achieved. However, misuse leads to a very severe limitation of the technique. In the United Kingdom most colour contrast penetrant testing is carried out using the solvent removal technique.

Misuse of the technique, which is depressingly commonplace, has given penetrant testing an undeservedly bad name among engineers. Missed defects have an unpleasant habit of revealing themselves at embarrassing moments.

In manual techniques solvent removers must *only* be applied using the wipe method. Dry clean cloth or absorbent paper is used to wipe away the bulk excess, and then fresh clean cloth or absorbent paper is wetted (*not* soaked) with solvent remover and used to wipe the surface clean. The surface is checked for background either in good light (500 lux) in the case of a colour contrast penetrant system or under UVA (black light) in the case of a fluorescent penetrant system, and the wiping process is repeated if necessary. All other methods of applying a solvent remover, such as spraying, flooding or immersion, will (not might) cause severe loss of sensitivity and are therefore forbidden. It is no use complaining that the surface cannot be cleaned unless one of the forbidden methods is used. If this is the case the wrong penetrant process has been selected and this error should be corrected.

It is unusual to find solvent removers in use in automatic penetrant lines but it does happen. In such a case the solvent is applied as vapour and allowed to condense onto the part and wash the penetrant off in the same way as oily contaminants are removed using a vapour degreaser. This method can only be used in automated lines. Apart from the technical dangers to the penetrant process, the health and safety aspect of the manual operation of such a process should preclude it. When such vapour phase removal is used care must be taken to ensure that the condensing solvent is free to flow from the surfaces of the parts and cannot collect anywhere to form reservoirs of liquid and to redeposit penetrant. The timing of such an operation is obviously critical and must be set up very carefully before a line is used. It is possible to show that with this technique a removal time of 20–30 s leaves excellent indications but after 40 s they are lost. It is necessary to remove the solvent from time to time and redistil it since contamination by the penetrant builds up to such an extent that the penetrant co-distils with the solvent and the result is very streaky background at inspection.

5.6 APPLICATION OF LIPOPHILIC EMULSIFIERS

The use of lipophilic emulsifiers has become uncommon in modern penetrant testing.

Lipophilic emulsifiers are generally applied by immersion followed by drainage; spray application followed by drainage has also been successfully used. When lipophilic emulsifiers are used lipophilic emulsifiers penetrant and emulsifier diffuse into each other, thus making the surface excess penetrant water washable. It is necessary to stop this action before penetrant entrapped in discontinuities becomes affected by the remover. While parts coated in penetrant are immersed in the emulsifier the process of mutual diffusion is quite slow. When the parts are withdrawn from the emulsifier it flows over the surface and the diffusion process is speeded up by the exposure of the penetrant layer to fresh emulsifier as the flow proceeds. This behaviour results in irregular removal of penetrant unless care is taken to turn the part so that the emulsifier flows regularly across the entire surface. This may be necessary even for fairly simple shapes. Figure 5.4 shows a cylindrical part. If such a part is processed using a lipophilic emulsifier and it is not turned during the emulsification drainage period, the top may have excessive background due to under-removal, the bottom may be ideally processed and the sides where emulsifier passes at greatest speed may have suffered over-removal. When emulsifier times are very short, e.g. 45 s, and a long part is being processed, the actual time spent in immersing the part in the emulsifier and then withdrawing it can reduce the emulsifier contact time achieved at the top with respect to the bottom by as much as 10%, thus decreasing the uniformity of the process. The hydrophilic remover process does not suffer from this problem and this is why it has steadily replaced the lipophilic emulsifier system over the years since 1958 when it was introduced.

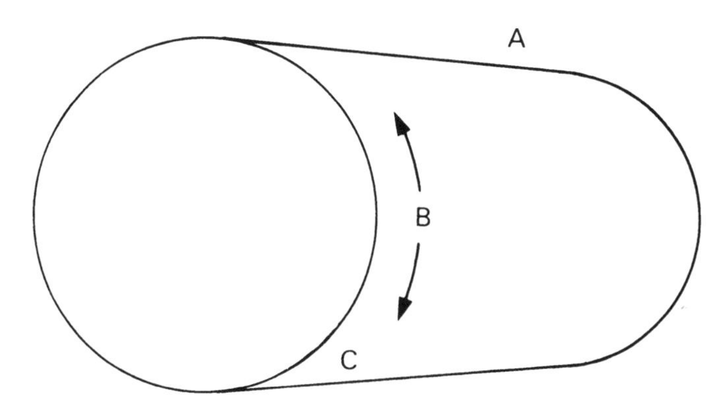

Fig. 5.4 When round components are processed by way of the post emulsifiable penetrant process, using a lipophilia emulsifier; the contact time giving proper removal at B, will not be the same as that for similar performance at A or C.

Lipophilic emulsifiers can be applied by flowing them on to the surface and allowing them to flow across the penetrant covered surfaced. Under *no* circumstances should emulsifiers be applied by brush. This method will affect penetrant trapped in defects and cause loss of sensitivity.

5.7 SPECIAL REMOVAL TECHNIQUES

A number of special penetrant removal techniques have been developed, usually to meet special circumstances, and undoubtedly further ingenious processes will be devised.

5.7.1 Foam removers

The use of detergent solutions in the form of foam is probably the most common of the special removal techniques. This technique was devised to overcome the difficulties of removal of surface excess penetrant from the insides of bores. The foam itself exerts detergent action and this is assisted by the physical force released by the bubbles as they break against the surface.

When foam removal is applied in an installation an arrangement similar to that shown in Fig. 5.3 often used. Low concentrations of detergent (hydrophilic remover) in water, normally below 5%, are used as these tend to foam well and the lower concentration allows better control of the activity of the remover. Foam can be sprayed on in some instances, and this has the attraction in the post-removal hydrophilic process that a pre-rinse is not necessary.

It is quite possible that foam removers will play a major part in penetrant testing in the future. The use of volatile solvents in all industrial processes is now under review from many aspects. They tend to be costly, offer health and safety problems, as they are either flammable or possibly harmful, and their possible effect on the environment is under close scrutiny. Foam removers can be provided in aerosol tins and some of the less suspect propellants can be used. The technique of using the penetrant process will in most cases need changing from

penetrant – solvent remover – developer

to

penetrant – foam remover – water wash – dry – developer

which may seem more complicated for *in situ* inspection where the colour contrast process is used. However, the removal of one of the processes using a solvent could be seen as a great benefit. If an aqueous developer were used and the procedure then changed to

penetrant – foam remover – water wash – developer – dry

before inspection the need for a volatile solvent would disappear. The present concerns over health and safety and over environmental problems may promote interest in such processes and acceptance of the practical changes involved.

5.7.2. Rolling in corn husks

Some components are fragile or extremely vulnerable to corrosion. In such cases the process of immersing the penetrant-coated components in corn husks or even chopped straw which absorbs the surface excess penetrant has been used. This process normally requires a water rinse and so can only be applied to components which are small enough to be dried very rapidly if corrosion is the problem.

5.7.3 Use of fruitstone grit or sand

Various fruitstones including plum and cherry, can be reduced to a very absorbent grit. Nutshells are also suitable. These materials can be sprayed onto surfaces which are covered with penetrant. The organic grit absorbs the surface excess penetrant and can be removed using an air blast. Fine sand will also absorb penetrant and act in a similar way. These techniques have been used quite successfully for cases when components must not be in contact with water or suitable solvents. However, they are difficult to operate satisfactorily and the removal medium becomes saturated with penetrant quite quickly so that frequent changes are necessary. Modern corrosion preventatives have reduced the need for this special technique.

5.7.4 Ultrasonically agitated solvent

Normally immersion of components in a solvent bath for removal of surface excess penetrant is prohibited without the added activity of ultrasonic agitation. However, application of penetrant to the rough ceramic coatings used as thermal barriers on modern turbine components requires exceptional removal techniques. One of these involves immersion in ultrasonically agitated solvent. This application is still under development and controls such as removal time, temperature range etc. are not yet established.

5.8 DRYING SURFACES

The stage at which surfaces are dried depends on the type of developer selected. When a non-aqueous developer such as a dry powder or a non-aqueous suspension is used, surfaces must be dried thoroughly directly after the final water wash and before the developer is applied. However, if an aqueous developer is used, either in solution or in suspension, it can be applied directly after the final wash. Aqueous wet developers can be applied to dry surfaces but they must be dried afterwards.

Whether drying precedes or follows application of developer the same rules apply: not too high a temperature and not for too long. The amount of penetrant viewed at a defect site is extremely small and heating it to temperatures of over 100 °C for half an hour does not increase its chances of being seen. An ideal to aim for is that the part surface should reach a temperature of around 60 °C in moving air and should be kept there long enough to drive off residual water. The applied temperature and the duration of drying obviously depend on the mass of the part. A part which is effectively a mass of stainless steel weighing 300 kg is going to require considerably more heat than a thin piece of titanium weighing a few grams. Actual drying times should be established experimentally and kept to in production. There are a number of aids to drying. One is to use a hot dip which consists in literally dipping the part in water at around 80 °C – the whole operation takes 20–40 s. This heats the surface of the part making drying very efficient. It might be thought that such a process is dangerous to sensitivity but this is not the case. Non-water-washable penetrants are hydrophobic and so repel both hot and cold water, and water-washable penetrants contain emulsifying agents which tend to thicken to a gel in hot water so that the indication becomes more firmly fixed in the discontinuity.

It is also very helpful to have available a supply of dry compressed air at low pressure (0.3 bar or 5 p.s.i.) to blow water out of holes and keyways and to remove any large droplets before the part enters the drier. It is *not* a good idea to have an air blast in a closed cabinet on an automatic penetrant line as this can create a mix of penetrant washings which can cause background problems (this is only in the case in water-washable penetrant processes).

Drying is carried out using mobile warm air either in a recirculating drier cabinet or as a blast. Static heat is very inefficient. In localized spot check systems a hand-held warm air blower can be used to very good effect.

5.9 APPLICATION OF DEVELOPERS

Penetrant developers come in four widely known types. Two of these are very commonly used, one is used widely in some industries and the fourth is becoming obsolete. There is a fifth type of developer which is used for special purposes.

5.9.1 Dry powder developer

Dry powder developers are light fluffy powders with a bulk density of around 0.15 g/cm^3. Dry powers are non-aqueous developers and must be applied to a thoroughly dry surface as any moisture will lead to a confusing situation during inspection. Unless they are electrostatically charged dry powders will not stick to dry surfaces, they only stick to the areas where penetrant has exuded back out of entrapment if the surface is properly dry. For this reason it is almost impossible to apply too much dry powder developer. This is a major reason why aerospace companies favour this form of developer.

Dry powder developers can be applied by powder storm in a closed cabinet, electrostatic spray gun or from a hand-held powder puffer. They are frequently applied by immersion and attempts have been made to use a fluidized bed.

The best method for applying dry powders is probably the electrostatic spray. This method has the very great advantage that new powder is used in each application. These developers are not simply a white powder but are a carefully selected blend of materials designed to contribute to the sensitivity of the system. Any method of using these powders which causes the mixture to become disproportionated or allows it to become contaminated is best avoided. Powder storm cabinets are very commonly used and are very successful provided that care is taken to control the station carefully. Even with a good extract and return system to protect the powder, on any fairly busy line the developer should be changed every fortnight.

A powder puffer can be used for local application. Some installations have only a tank into which dry powder developer is emptied. The best way of using this is to mount the parts clear of the developer and dust the powder on. Immersion of parts will lead to contamination of the powder leaving flecks of fluorescence in it which can become deposited on fresh parts thus giving rise to false indications or confusion at the inspection stage. Open tanks full of developer will also become damp, thus causing loss of sensitivity on the part of the developer. Developer powder in such arrangements needs to be changed weekly on a busy line. The possibility of using a fluidized bed to apply dry powder developers has a number of attractions. Attempts to do this have been only partially successful because the bulk density of the powder is very low. In this case control of the powder is very difficult.

5.9.2 Non-aqueous suspension developers

Non-aqueous suspension developers are frequently used, particularly in colour contrast penetrant systems, and are widely abused. Properly used this type of developer can contribute the greatest sensitivity to a penetrant process. Since the non-aqueous developers consist of inert developer particles in a volatile organic solvent, they are unique among developers in that the solvent invades the defect, dissolves some of the penetrant and draws the penetrant into the developer layer to be displayed for inspection. The common problem is that *far too much* developer is applied. If defects are small or penetrant removal has been too severe, thus reducing the amount available to be developed, it is quite easy to mask the indication. The correct procedure for applying this type of developer is to provide a light even spray. In the case of colour contrast penetrant processes the developer layer should be just thick enough to provide an even white background, and in the case of fluorescent penetrant systems it sould be such that it is possible to see the test surface through the developer layer. Non-aqueous-suspension developers can be applied *only* by spray techniques after thorough agitation. All other methods of application are *forbidden* since they will cause loss of sensitivity because the solvent will wash penetrant from defects before the developer has a chance to work.

5.9.3. Aqueous solutions

Aqueous solutions are the newest type of developer, having been introduced about

25 years ago. They consist of blends of water-soluble chemicals which form a developer layer when dried out. These developers have found favour in the automotive and aluminium industries. They tend to provide very clearly defined indications which show little tendency to overbleed, and stable indications on samples can be kept for several days. When such a developer is used as part of a colour contrast system it must be soluble enough for the concentration of developer to be sufficient, when dried out, to provide a continuous pale background.

Application can be by immersion, wet spray, flow on or brush on. Parts which have complex shapes should be turned to drain any large entrapment of the solution. Aqueous solution developers can be applied to wet or dry surfaces. However, the layer of developer solution *must* be dried actively after application, otherwise indications will be blurred and indistinct if they appear at all. The only points to watch for in the case of aqueous solutions are control of concentration – water is quite volatile – and control of contamination.

5.9.4 Aqueous suspensions

Aqueous suspensions are mixtures of inert developer powders in water. As in the case of non-aqueous suspensions the developer must be agitated thoroughly before application. Application can be by spray or, more commonly, by immersion in an agitated bath. In each case the developer can be applied to either wet or dry surfaces and, as in the case of the aqueous solution developers and for the same reasons, the layer applied must be actively dried – passive drying is useless. Drying is achieved by circulating warm air at a temperature between 70 and 75 °C as in the case of aqueous solution developers.

The major difficulties with this type of developer are as follows.

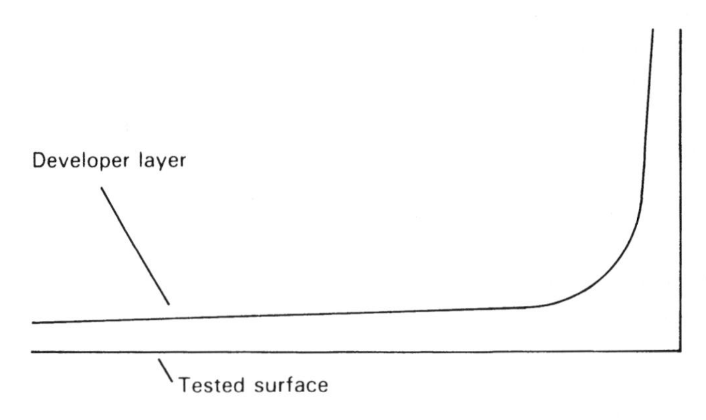

Fig. 5.5 Aqueous suspension: developers leave an irregular layer particularly at changes of section.

1. It is difficult to achieve a uniform layer where the parts have abrupt changes of section (Fig 5.5).
2. It is difficult to maintain proper control over the concentration and proportion of developer solids in the suspension.

5.9.5 Lacquer developers

Lacquer developers can be considered as special forms of non-aqueous suspensions. They are in fact non-aqueous suspensions with a resin dissolved in the solvent. The purpose of the resin is to freeze the indications as the developer dries out and to provide a peelable layer which can be stripped off and stored as a record of the indication.

Such developers are frequently used on test pieces to provide a record of the defects in them so that a rapid functional test can be carried out on a line to ensure that it is working. However, it should be remembered that lacquer developers are very sensitive and provide very good resolution, and this fact must be taken into consideration if a transfer is stored and used for comparison with a penetrant system which uses a different type of developer, e.g. a dry powder.

5.10 VIEWING CONDITIONS, INSPECTION AND INTERPRETATION

Although it is now possible to automate the viewing of parts by using conventional optical systems or laser scanners, and the image can be digitized and compared with a pre-set acceptance standard for an accept–reject process and that data stored, such installations are expensive and as the variety of parts to be inspected increases so does the complexity of the system and its cost. Some automated systems for the inspection of high volumes of parts of the same or similar shape are available, but they are very expensive.

In the view of the difficulty and cost of non-human visual inspection it is probable that penetrant indications will be examined by eye for a long time to come. The conditions of inspection vary depending on whether a colour contrast system or a fluorescent system is used.

In the case of colour contrast systems the penetrants are always red. This is because red on white provides the best coloured contrast for the human eye. Fortunately for designers of penetrant systems human eye response is universal for over 92% of the world population. The remaining 8% represent those people, nearly all male, who suffer from defective colour vision. It is not a good idea for people who do not have normal colour vision to be involved in inspection of penetrant indications. The choice of red is useful in that in daylight or under tungsten lamps a true red is seen. Under the ubiquitous low pressure mercury arc strip lights the red will darken. Colour is one of those phenomena which is recognized by all people, except those with defective colour vision, but is difficult to define. The quality of colour depends, among other things, on the quality of the light falling on it. The common strip light is very low in the type of light which shows reds to good advantage. This is why the effect is 'cold' and also why butchers and greengrocers avoid their use or use them with great care. Who wants to buy grey meat or tomatoes? In a factory using strip lights red indications will appear darker, thus increasing contrast. Inspection of fluorescent indications must involve excitation of dyes in the penetrant and the results depend on the power of the illuminating UVA lamp, the fluorescent efficiency of the dyes used and the level of contrast. The choice of dyes is made by the manufacturer. The level of UVA illumination depends on the lamp supplied and should be checked regularly on a monthly basis. Minimum output should be around 1000 W/cm^2 (this figure varies from one specification to another) and a good level of darkness should be maintained (less than 10 lux of white light is acceptable). It is best that inspectors use dark overalls since white ones are washed with mixtures containing fluorescent additives to achieve the 'whiter than white' look. In the presence of UVA lamps these white coats will glow white and greatly increase the level of white light in the inspection booth, thus making inspection more difficult than necessary.

It must be remembered that what is viewed is colour, not fluorescence. The colour is the result of the process of fluorescence. Green is chosen as a colour for fluorescent penetrants because human eyes see green very well in dark conditions. In normal daylight yellows and reds are the easiest colours to see. As daylight fades the highest sensitivity of our eyes to colour shifts from yellow to green. This is known as the Purkinje shift after the Czech scientist who discovered this. The inside of an inspection booth must be kept tidy and free from reflective or light-reflecting objects. Inspection should be carried out in a deliberate and orderly way.

The inspector must also be examined. We have already noted that people with defective colour vision should not become inspectors for penetrant testing. All uncorrected defects of vision are undesirable for inspectors, and an inspector should be tested for satisfactory visual function every 6 or 12 months.

Inspectors must also take into account the fact that our eyes change as we move from a light to a dark environment – this is called dark adaptation and is not an instantaneous change. Inspectors should allow 10 min for their eyes to become dark adapted before they start looking for indications. Inspection is a job which requires great concentration and so there should be minimal interruption from outside and the booth should be sited in a quiet area away from unnecessary noise. Concentration and visual acuity also start to decline if an inspector works for too long without a break. A short break of around 10 min needs to be taken after every 2 hours work. The inspector should leave the dark booth for this period and go into a white light area. Inspectors must remember to wait to become dark adapted again after such breaks before starting to inspect parts.

Inspection and interpretation are two quite different operations which are frequently confused. Inspection involves looking at processed parts and recognizing and recording the fact that indications are or are not present. Interpretation involves deciding what caused the indication. Deciding whether an indication is actually relevant or not is an important aspect of interpretation. The most common source of irrelevant or false indications is poor processing – poor wash, dirty developer, operators with dirty hands etc. Touch marks where parts are jigged or come into contact with each other during processing can be a source of irrelevant indications.

Relevant or interesting indications fall into a number of categories according to their appearance. Continuous lines, jagged lines and intermittent lines are normally signs of cracks, forging laps, scratches, cold shuts or die marks. Intermittent lines should always be regarded with the greatest suspicion. Cases are known where two short linear indications 15 cm apart were the only signs of a crack extending the full length between them. The undetected part was filled with contaminant. Any part showing short cracks should be examined visually at a magnification of 5 × to determine which there are more widespread problems. Dots and rounded indications can be signs of porosity, blowholes or corrosion pitting. A considerable amount of experience is needed before accurate interpretations of indications can be made.

5.11 POST-CLEANING

After carrying out penetrant processing it is normal practice and a wise precaution to clean up surfaces. After exposure to the atmosphere developer layers may become hygroscopic, attract water and help to initiate corrosion. Residues from some developers are soluble in water and can be washed away readily. Others leave a residue which must be wetted and then rubbed away using a brush or cloth. Water-based removal processes remove only the developer residues and leave any penetrant residues untouched. The method for removal of penetrant residues is to vapour degrease the parts or solvent wash by spray or immersion. Residues which prove difficult and resist this type of removal can be boiled in high concentration detergent solution (30%) for 30 min.

Clean-up processes are often followed by application of a protective.

5.12 SPECIAL TECHNIQUES

Not unnaturally there are special circumstances requiring materials and techniques not covered by the major systems described so far. The problem of penetrant removal has been solved by rolling parts in corn husks, spraying with fruitstone grit and washing off and a number of other esoteric methods. Other special materials and techniques are listed here.

5.12.1 Low temperature materials

The processes of penetrant testing are chemical, physical or both and, as such, are influenced by temperature. As temperature falls so the rate of chemical and physical processes falls. A rough guide is that the time taken for a process will double for every fall of 10 °C. This means that materials designed to work at 20–30 °C do not work at temperatures below zero. Special products which remain stable liquids at temperatures of −30 °C are required. Clearly, water is useless at these temperatures. Even with special products the process is much slower at low temperatures. Minimum penetrant contact time for penetrants is 30 min and developers also need 30 min to work. Special low temperature materials are available for colour contrast or fluorescent penetrant testing between 5 and −30 °C.

5.12.2 High temperature materials

Just as low temperature materials are required so are high temperature materials. In this case it is not the variations in weather which are the cause of the requirement. Regular penetrant materials can be persuaded to work at temperatures between 10 and 70 °C which covers the hottest situation likely to be encountered as a result of weather or climate. High temperature penetrants are designed to work at temperatures up to 180 °C. Such materials tend to be very viscous at ambient temperatures and become thinner at the working temperature. Application of both the penetrant and the remover is by swabbing on and the technique is manual. Because of the temperature only colour contrast penetrants can be formulated for high temperature use.

5.12.3 High resolution penetrant systems

High resolution of indications can be achieved using lacquer developers (section 5.9.5). This process can be further refined by using special penetrants which ensure a very high level of dye concentration. After application of the special penetrant and the elapse of its contact time the best method of removal is by hand rubbing a foam of hydrophilic remover. Lacquer developer is then applied to the dry surface. High resolution systems are of great help when a local check with high sensitivity is required.

5.12.4 Thixotropic penetrant materials

Penetrants are particularly mobile liquids which are designed to spread and wet very thoroughly. There are occasions when it is helpful for the penetrant to be restricted to a certain area. One attempt to solve this problem is to use a thixotropic penetrant. If the thixotropic gel is broken down at the time of application the process works quite well; otherwise, results can be very variable.

5.12.5 Reverse fluorescent penetrant processes

From time to time interest is shown in what is known as reverse fluorescence penetrant testing. In this case the developer is fluorescent and the penetrant absorbs UVA. At inspection the indications appear black against a fluorescent white ground. Very good sensitivity can be achieved in the laboratory. However, it is difficult to identify any advantage over conventional fluorescent penetrant testing and there are some disadvantages, notably that the whole background must be coated evenly with developer and any incompleteness of cover will appear bluish black under UVA and lead to confusion at inspection. It is equally difficult to see any advantage over colour contrast penetrant testing. Certainly higher sensitivity can be achieved, but this is true of conventional fluorescent penetrant testing. One major reason for using colour contrast penetrant testing is that there is no need for a UVA lamp and so no need for electrical power. A second reason is that there is no need for the inspection stage to be carried out in darkness. Application of materials for the reverse fluorescent process is identical with that for conventional materials.

5.12.6 Water-based penetrants

Some materials react with or are dissolved by hydrocarbon-based chemicals. Special water-based penetrants exist for this purpose. They are used on plastics and on systems which will later be exposed to liquid oxygen. It is not normal for certificates to be issued for liquid oxygen compatibility as tests are very specialized. However, water-based penetrants are used on such systems. Such penetrants are water washable and, of course, can be very easily over-washed. Special care must be taken to reduce over-removal and wipe techniques are favoured where possible.

QUESTIONS

1. Components need cleaning and drying before penetrant testing. Which of the following methods of preparation are best avoided?

   (a) Grit blasting (mineral)
   (b) Vapour degreasing
   (c) Fruitstone grit blasting
   (d) Detergent cleaning

2. Some commonly used cleaning chemicals can interfere with brilliance of fluorescent penetrants to the extent that in some cases it is destroyed. Which of the following cleaning chemicals must be washed away completely before penetrant testing with a fluorescent process so that this problem is avoided?

(a) Acid ferric chloride
(b) Nitric acid
(c) Acid chromates
(d) All of these

3. When penetrant is applied by electrostatic spray it is necessary to inspect the components under UVA (black light) before passing them to the penetrant contact stage. The reason for this is

   (a) If this is done quickly enough some indication of where defects occur can be noticed
   (b) to ensure that over-spray is minimal
   (c) to check that the entire surface has been covered
   (d) to estimate the necessary remover time

4. The major disadvantage of an immersion water wash in the water-washable penetrant process is that

   (a) the temperature of the water is very critical and may influence the sensitivity of the process
   (b) too much water is used and a large effluent plant is needed
   (c) the water becomes progressively contaminated and control of the process is reduced
   (d) the water bath starts to foam as components are immersed

5. When detergent (hydrophilic) remover is applied by immersion of the components a water pre-rinse is normally used. This gives the advantage that

   (a) a thin regular layer of penetrant is left on the surface of the components, thus ensuring uniform removal
   (b) the detergent (hydrophilic) remover solution will not tolerate even quite low concentrations of penetrant and become spent rapidly if the pre-rinse is avoided
   (c) the pre-rinse can remove 95% of the surface excess penetrant and gives a mixture which can be separated readily thus allowing effluent control to be managed readily
   (d) all these points are advantages to using a pre-rinse

6. Electrostatic application of dry powder developers is advantageous because

   (a) without an electrostatic charge dry powders will not stick to dry surfaces and so a thick layer can be applied
   (b) less powder is used and so it is more economical
   (c) new powder is used at each application so that the problem of loss of smaller particles is avoided
   (d) the powder remains drier than is the case with other methods of application

7. Dry powder developer is preferred by many aerospace companies because

   (a) it only sticks to wetted areas of the surface such as are created by penetrant bleeding back from entrapment in defects
   (b) it offers the greatest contribution to sensitivity
   (c) it is the most economical developer
   (d) it is easily used in automatic penetrant installations

8. The commonest error made when non-aqueous suspension developers are used is that

   (a) operators forget to agitate the suspension before applying it
   (b) far too much developer is applied
   (c) operators do not allow time for the developer to work before inspection
   (d) application is often by other methods than spraying

9. Colour contrast penetrants are universally some shade of red. The choice of red is established because

   (a) the human eye is very sensitive to the contrast of red on a white background in daylight
   (b) the original specifications were written when only red dyes were stable enough to use and they have not been changed
   (c) red dyes cost less than other colours
   (d) dyes of other colours are very difficult to dissolve in a suitable base for a penetrant

10. The Purkinje shift is

    (a) a defice for moving baskets of components from one station to another in a penetrant installation
    (b) the change in wavelength from absorption of UVA (black light) at 365 nm to the emission of a yellow-green colour at 520–530 nm which is seen in fluorescence
    (c) the change in colour sensitivity of the eye as the white light level is reduced to darkness
    (d) the name given to the time that inspectors must wait on going into an inspection booth from normal white light to allow their eyes to become dark adapted

Note: There is only *one* correct answer to questions 3–10. Tick your choice and check it with the correct answers on p. 213.

11. Describe the serious disadvantages of mechanical pre-cleaning methods for preparing components for penetrant processing and suggest a method for overcoming this difficulty if such methods cannot be avoided.
12. Provided that penetrant reaches all surfaces of a component any method of application is acceptable. What consideration must be taken into

account if the results of the test are not to vary depending on the choice of method of application?

13. Water washing is used in four different stages of commonly used penetrant processes. What are they and what conditons should be controlled to ensure that the process is within specification and is repeatable?
14. What is the commonest method for application of lipophilic emulsifiers and what special steps must be taken to control this type of removal?
15. Describe the conditions for drying surfaces which should be used to ensure good processing and good results.
16. Etching of components before penetrant testing has the advantages that the edges of defects are opened up and any oxide or contaminant is removed. In view of these advantages why is it not used for preparation of all components?
17. Electrostatic application of penetrant is very advantageous in two specific situations. (It is also useful in many others.) What are the two very specific instances and why is this type of application so helpful then?
18. Solvent removers are widely used in colour contrast penetrant testing. Why must these removers be applied by the wipe technique only when this process is operated manually?
19. When components are inspected after a fluorescent penetrant process they must be viewed under a UVA (black light) source of a minimum intensity in a darkened area with very low white light levels. Explain why these conditions are necessary.
20. Explain briefly where thixotropic penetrants might be helpful and describe any special aspects of their application.

6 Choosing a penetrant system

Straightforward consideration of the basic material available for penetrant testing, i.e. the penetrants themselves, their removers and their developers, allows an impressive number of combinations, and when the various levels of potential sensitivity and methods of applying materials are accounted for there are more than 1000 possible ways of using this basically simple technique. Possibly the most surprising aspect of this enormous choice is that so few of the methods are redundant or of solely academic interest.

In this chapter we consider seven factors which must be taken into account in selecting a penetrant system:

1. specification requirements
2. part considerations
3. sensitivity requirements
4. safety factors
5. production considerations
6. ecology
7. economy

6.1 SPECIFICATION REQUIREMENTS

When a penetrant process is governed by a specification, much depends on the actual specification. Some penetrant processing specifications are very vague, others are fairly specific and some are very specific indeed. National specifications tend to be very vague, whereas professional society specifications such as those of the American Society of Mechanical Engineers (ASME) tend to be more specific. Manufacturers' specifications such as those given by aerospace manufacturers tend to be very specific indeed.

When working to a specification many of the points of choice have already been made by the specifying authority. The choices of a colour contrast or fluorescent process, a water-washable or post-removable process, the type of developer and even the potential sensitivity of the method will be set out in detail. Nevertheless there is a wide range of choice which must be made within the requirements of any specification if penetrant testing processing is to be successful.

6.2 PART CONSIDERATIONS

6.2.1 Size considerations

Table 6.1 summarizes the influence of size. Enormous parts such as aircraft wing spars are best processed by spray application of penetrant in a post-removable or water-washable system. Electrostatic spraying is also very attractive.

Table 6.1 Part considerations for penetrant applications: size

Size	*Appropriate systems*			
	Post-removable		*Post-emulsifiable*	*Water-washable*
	Spray	*Immersion*		
Enormous (wing spars)	×			×
Large (processed on hoist)	×	×		×
Normal (in baskets)		×		×
Small (fasteners, bolts)		×	×	×

When the shape of these parts is simple, the final choice will be made on considerations of surface finish and sensitivity requirements. Large parts which require individual processing on a hoist such as turbine discs and large castings can be sprayed with penetrant or immersed. In some cases electrostatic spraying is the most useful whereas penetrant is often best applied to complicated large parts by conventional spraying. Normal sized parts can be processed in baskets or on carriers using any appropriate method. Immersion application of penetrant is normally used in these instances. However, if such parts have small holes such as cooling holes in turbine blades they are best jigged and the penetrant applied by electrostatic spray. The Faraday cage effect in electrostatic spray systems prevents penetrant from entering these holes and causing insoluble removal problems with consequent excessive bleed-out during development and inspection. Small parts such as nuts, bolts and fasteners present special problems, particularly along threads, and care must be taken to account for this.

6.2.2 Surface finish

Surface finish is one of the most important and even limiting factors in penetrant testing (Table 6.2). Surface finish has the most profound influence on

Table 6.2 Part considerations for penetrant applications: surface finish

Size	*Appropriate systems*			
	Post-removable		*Post-emulsifiable*	*Water-washable*
	Spray	*Immersion*		
Smooth (sheet metal machine parts plated, polished)	×	×	×	×
Coarse (machined or cast surfaces)		×	×	×
Very coarse rough castings		×		×
Special coating (powdered metal)		×		
Materials which have been used in hot areas; service jet pipes etc.		×	×	×

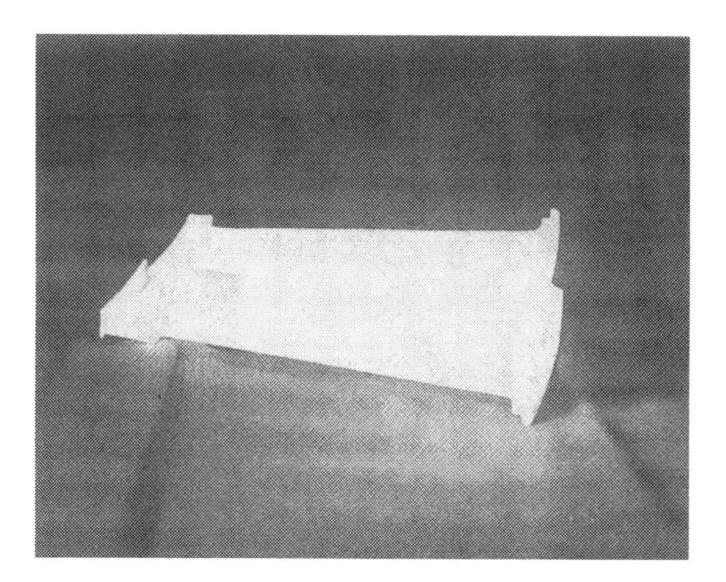

Fig. 6.1 Difficult surfaces: at first glance it might be thought that this aluminium stator segment had been processed in too brilliant a penetrant. However, this is not so. The surface is heavily corroded and the pits are easily identified through the background fluorescence.

removability, which is one of the major factors affecting sensitivity. In Fig. 6.1 an extreme example of the effect of surface finish on penetrant testing is shown. The component is heavily corroded and background fluorescence is strong over the entire surface, however, careful selection of a process and proper application allow effective inspection with indications of defects showing through the background. It is possible to process smooth parts in any of the common processes. Water-washable systems cannot be expected to give the levels of sensitivity achievable with post-removable systems.

Coarser surfaces including smoother castings and rougher machined parts or extrusions can be treated using any of the systems. Water-washable materials are generally more useful with very coarse surfaces, but post-removable systems can give very good results on surprisingly coarse surfaces and are very useful on porous surfaces. Powdered metals and special coatings are best dealt with by the post-removable systems. When parts have been used, particularly if they have been subjected to great heat, they may require special attention at the penetrant removal stage.

6.2.3 Conformation and shape

Conformation and shape are closely related. Solid parts with simple geometry can be processed using any of the available systems (Table 6.3). Post-emulsifiable systems, although not post-removable ones, should be avoided with hollow parts or when geometry is complex. Spray application of hollow parts is not straightforward, but electrostatic spray systems can be designed to give optimum results with some of these.

Table 6.3 Part considerations for penetrant applications: conformation and shape

Conformation and shape	*Appropriate systems*			
	Post-removable		*Post-emulsifiable*	*Water-washable*
	Spray	*Immersion*		
Solid	×	×	×	×
Hollow	ES	×		×
Simple shape	×	×	×	×
Complex shape	×	×		×

ES, electrostatic spray

6.3. SENSITIVITY CONSIDERATION

When people unfamiliar with penetrant testing first come into contact with it they frequently require to be shown 'everything'. It becomes apparent very quickly that such an approach is unhelpful. The very nature of metallic surfaces at an atomic level ensures that some dislocations and defects are present. It is relevance which is important. For example, some degree of porosity is inevitable in aluminium alloy castings, yet these are quite acceptable and do not impair the usefulness of the part.

Sensitivity considerations involve the design of systems which will find defects which matter. The acceptability of discontinuities must be laid down by design engineers. Penetrant testing can demonstrate defects but cannot

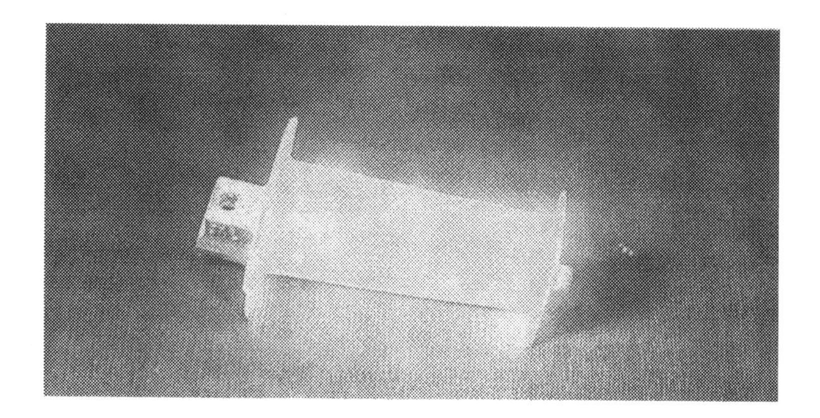

Fig 6.2 The possible effect of choosing the wrong penetrant process. This blade has been processed with a penetrant of far too high a level of fluorescent brilliance. The result is a surface which is almost totally fluorescent and it is impossible to see any indication under this. Any removal process capable of removing the background fluorescence would also remove penetrant from defects.

make judgements on them. It is possible to make up a league table of penetrants based on their performance on specified test plaques or parts. Before thinking in such a way or using the results of such an exercise it is as well to consider what is meant by sensitivity. In its simplest definition, as far as penetrant testing is concerned, it is the ability to find discontinuities. An example of the dangers of the league table approach to choosing a penetrant system is shown in Figure 6.2. Here a turbine blade has been processed by use of a very bright fluorescent penetrant which, in suitable applications, offers very high sensitivity. In this application the sensitivity is lost because the background fluorescence is as bright as that which is due to indications of defects which may be present.

For a process to have sensitivity the penetrant itself must be able to cover surfaces and invade discontinuities, the surface excess must be removable without removing the material entrapped in discontinuities, the penetrant must emerge from discontinuities readily after surface excess has been removed and finally it must be possible to see the penetrant. It must not be forgotten that developers also contribute to the sensitivity of penetrant processes. Not only do the different types of developer contribute their own level of sensitivity to the system but there are wide variations from product to product within the developer type.

Sensitivity levels for the whole system should be established experimentally for each specific application. Specifying authorities will conduct experiments to build up knowledge of the performance of penetrant systems and to classify them on the basis of this experience. Two points must be remembered when using data obtained using this approach: the first is that the classification is *not* a league table but an assessment of fitness for purpose, and the second is that the classification is designed around a specific range of parts or structures.

6.4 SAFETY FACTORS

Safety factors are clearly very important. Testing facilities have been known to burn down through failure to take full account of flammability of materials. Materials with flash points of 100 °C and above can be regarded as acceptable. Materials with flash points below 40 °C should be used with the greatest care and only in special circumstances. Use of low flash materials in bulk is best avoided.

The volatility of penetrant materials is also important. Highly volatile materials tend to be disliked by inspectors and operators, and should only be used in solvent wipe systems in local checks. At best volatile chemicals are a nuisance to operators and impair competence and efficiency, and at worst such materials offer a health hazard.

Dust produced by improper use of dry powder developers is a nuisance and interferes with operator efficiency. Prolonged exposure of the skin to penetrant materials can cause cracking and drying. The use of rubber or PVC gloves can overcome this very simply.

Until the mid 1990s the choice of volatile solvent as a penetrant remover and as a carrier liquid for non aqueous suspension developers was 1,1,1, trichloroethane. Since this material has become unavailable owing to environmental concerns and its related chlorinated hydrocarbons which remain available present unacceptable health and safety characteristics flammable solvents are now used. These present the well known risks of fire and explosion. These materials form explosive mixtures with air in concentrations in the range of 1 and 12% by volume. Consequently they must be used with suitable precautions. Avoidance of any form of ignition including static electrical discharges is essential. The use of such materials in enclosed spaces requires continuous monitoring of the local atmosphere and exchange of air.

6.5 PRODUCTION CONSIDERATIONS

The availability of equipment and its design are critical factors. Equipment may be automatic or manual. The operating parameters, penetrant contact times, washing or other removal requirements of either type can be varied. The availability of spray application has two significant advantages in that new material is used at each application and a wide range of different penetrants can be used on one line. In selecting a unit the size, weight and number of parts to be processed must be accommodated.

Special recommendations for production facilities include the following:

1. Low pressure (less than 0.35 kg/cm^2) air lines to blow off and spread out excess water prior to oven drying.
2. oven control settings which can be set and locked;
3. appropriate exhaust facilities;
4. proper viewing area with good white light or black light illumination as appropriate;
5. removal of penetrant materials.

6.6 ECOLOGICAL STATUS

The ecological status of all process materials is important. The effect of effluent must be taken into consideration. Penetrants are usually oily liquids which can be harmful to the environment. Emulsified rinsings are biodegradable but large quantities can cause difficulties. Post-emulsifiable systems are the worst offenders because they involve the largest quantity of material drag-out. Aqueous remover and water-washable systems are more readily dealt with and effluent from these can be cleaned up.

In designing a penetrant system the priorities must be as follows:

1. defects sought
2. surface finish
3. size and weight
4. safety
5. production facilities
6. environment and economy

6.7 ECONOMIC CONSIDERATIONS

Penetrating testing is one of the most economical methods of testing available but this cannot exempt it from economic measures. The following factors should be taken into account.

Each time a part is immersed in penetrant remover or developer or these materials are applied in any other way materials are used up. This is known as drag-out of materials. The most expensive process is probably the solvent wipe method using aerosols. The post-emulsifiable system is the most expensive of the bulk processing methods, and water-washable systems are generally the least expensive since they involve only two materials.

The number of processing steps influences the cost of processing parts. However, although the aqueous remover system has one more stage than the post-emulsifiable system the extra cost of material and material losses in the latter system outweighs the advantage of fewer stages.

Water can be a costly item in some parts of the world or when demineralized water is used. Wash water can be cleaned up by coalescers flocculant systems, ultrafiltration process, reverse osmosis, carbon filters and other methods, but this is an extra cost in itself. Much greater volumes of water are required for automatic plants than for manual plants. The amount of energy used in the form of electricity is normally small, although automatic lines can generally be expected to use more power than their manual counterparts.

QUESTIONS

1. The size of individual components has a great influence on the choice of how penetrant is applied to them: true/false.
2. Components with very rough surfaces are most likely to be tested using

   (a) a lipophilic removal process
   (b) solvent remover in the vapour phase
   (c) a water-washable penetrant process
   (d) a undiluted (i.e. full strength) detergent (hydrophilic) remover

3. When choosing a penetrant process for a component the choice will be

   (a) for the penetrant with the brightest fluorescence and the maximum possible sensitivity
   (b) the process which costs the least
   (c) the process which takes the least time
   (d) the process which gives acceptable sensitivity while avoiding heavy background fluorescence or colour

4. When handling all chemicals due care must be taken with regard to safety. The flash points of material give an indication of their flammability. Which flash point ranges will give the need for the greatest care?

   (a) 90–105 °C (205–220 °F)
   (b) below 40 °C (104 °F)
   (c) 55–70 °C (130–158 °F)
   (d) non-flammable

5. When taking account of production considerations which of the following points must be considered?

   (a) throughput of components
   (b) component size
   (c) component weight
   (d) all of these

6. Give the order of priorities of the factors to be considered in choosing a penetrant process:

surface finish	( )
size and weight	( )
economy	( )
defects sought	( )
safety	( )
environment	( )
production facilities	( )

Note: There is only *one* correct answer to questions 1–6. Tick your choice and check it with the correct answers on p. 215.

7

Equipment for the penetrant processes

The penetrant testing process is so flexible that equipment can vary from a few aerosol cans or small tins and brushes or swabs through various levels of sophistication to a fully robotized line complete with automatic inspection. All systems and equipment must be designed and operated to allow uniform controlled efficient operation, stage by stage, and to allow for regular periodic cleaning and renewal of the chemicals used.

Of course, the actual equipment used is governed by the type of penetrant process to be used. The size and complexity of the actual system and its accessories is largely determined by the number of components to be processed and the maximum speed at which they should be inspected. Penetrant inspection must not form a bottleneck in the processing of parts and the capacity of any line must be designed around the *maximum* throughput required. There is little point in designing systems which will deal with the throughput calculated on the average required over a 12 month period if the number required to be processed each year must be inspected in four separate weeks of intense work. In such a case the rate of processing must be calculated on the annual production divided by the hours available in 4 weeks. The fact that the penetrant line remains idle for 48 weeks in the year is of minor consequence. The factors to be considered include the following:

1. the actual penetrant process;
2. the size of the largest parts passing through the line;
3. the peak number of parts passing through the line per hour;
4. the weight and mass of the heaviest and largest parts to be processed;
5. the variety of shapes, conformations, sizes and weights of parts to be processed on this line.

A quick reference to the flowcharts showing the stages of the various processes (Figs 4.4–4.7) indicates that the number of actual stations required depends on the process chosen. In most cases this selection will be fixed – particularly in the aerospace industry.

If the variety of parts passing through the line is very great it may be advantageous to combine spray and immersion techniques in the same line. This can be of particular interest on semi-automatic or fully automatic installations. It is also possible to combine various penetrant processes or various levels of sensitivity on the same line. In such cases, unless the line is automatically controlled, it is very important to ensure that the correct procedure is applied to each component.

The mass of the largest part to be processed is a further vital piece of information. The normal penetrant line is designed around a load of up to 100 kg. If this is exceeded, not only the handling equipment and strength of supports but also the heating arrangement in the drier station must be altered to deal with the increased load.

7.1 FIELD INSPECTION KITS

In many industries the use of field inspection kits, either purchased ready made or made up by the user, is the only practical way of carrying out penetrant inspection. The vast majority of colour contrast penetrant testing in the United Kingdom is carried out in this way. It is often necessary to inspect parts of large fixed structures which cannot be dismantled, and sometimes inspection takes place while such equipment is still in use. Even in the aerospace industry kits are useful for first-level inspection of such parts as undercarriages which are still fixed to the aircraft.

Such kits normally contain aerosol cans of penetrant, cleaner/remover and developer. A typical field inspection kit is shown in Figure 7.1. Non-aqueous suspension type developers are always supplied in aerosol cans in such kits as this is the only practical method of spraying the material. Such kits should also contain clean cloths or absorbent paper for the proper operation of the solvent-removal penetrant process. Some inspections require a water-washable penetrant process in which case a supply of water under moderate pressure from a hose with a spray nozzle is needed.

When fluorescent penetrant is used a UVA lamp (black light) giving sufficient energy (about 1500 $\mu W/cm^2$) at a distance of 30 cm is needed. Care must be taken to ensure good darkness for inspection when fluorescent penetrant testing is carried out in the field, and this is not always easy. Some situations are so difficult in this respect that night inspection must be considered.

Penetrants and removers for kits can also be supplied in small tins (typically 500 ml). This is most common in kits of special products such as those for high resolution, low temperature or high temperature. In such cases the penetrant can be applied by brush or swab. The best solution is for the lid of the can to contain a swab. In this case there will not be a loose brush in the kit which will coat all the other components with penetrant, and the likelihood of the swab being used for another material is reduced.

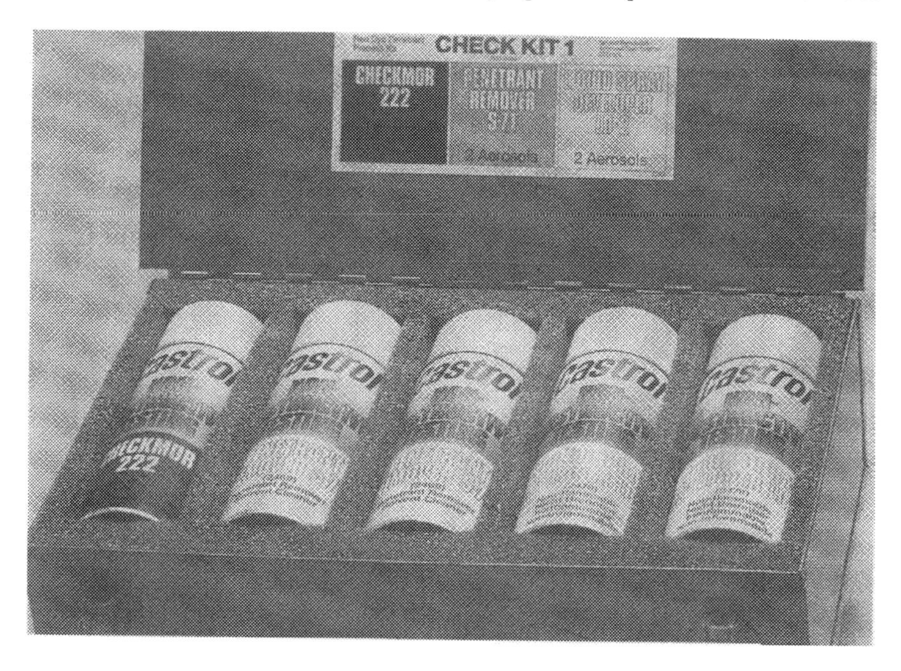

Fig. 7.1 Field test kits: the flexibility of the penetrant method of NDT allows the apparatus to be reduced to a portable case of aerosol cans.

7.2 MANUALLY OPERATED PENETRANT LINES

Penetrant lines which are purely manually operated are unusual except when only a few parts are processed at a time. Some form of handling of the parts is almost always included in the system. This is true of lines where the penetrant materials are applied by spray or by immersion. The physical problems of manipulation of baskets or individual parts through a sequence of wet and dry stations without some mechanical handling aid are self-evident, and the classification 'manual lines' includes those involving roller tracking with pneumatic raising and lowering of parts in some stages or the use of overhead cranes or turntables. Figure 7.2 shows a typical manual line.

Such manually operated lines can be self-contained or modular. The actual choice of line is governed by the inspection needs. If the number of parts to be inspected over a given time is small and their size warrants such an approach, a single spray booth can be very useful. If the parts are to be processed by a water-washable penetrant system the whole process from penetrant application through to inspection to be carried out in the booth. Manually operated spray can be electrostatic, conventional or a combination of the two types. Such a system is quite acceptable provided that it is possible to occupy the booth with a single part or load of parts for up to 90 min. The first refinement of such a system is for the inspection to be taken to another

Fig 7.2 A typical manually operated penetrant installation.

booth, thus releasing the processing station after 50 min. The next stage is to use a three-station system where penetrant is applied in the first station and washed off after the contact time has elapsed. Parts are then transported to a drier station, normally by way of an overhead conveyor, and then to a third station where developer is applied and inspection takes place. This concept can be developed further, when throughput justifies it, to the situation where a separate station is used for each stage.

When a post-removable penetrant system is used the stages of penetrant application and pre-rinse can be carried out in the spray booth and the part or parts removed to immersion stages for the application of remover and the first stage of post-rinse. Alternatively, the remover solution can be applied using a spray or foam gun and both this stage and the final water rinse can be carried out in the spray booth. As the number of parts to be processed per hour increases, it becomes necessary to increase the number of stages until there is a separate station for each step in the process.

In any spray system there is a need for extraction. Ideally extraction should be lateral and draw sprayed material across the part(s) and away from the operator. The force of such extraction must be enough to prevent problems for the operator but not so much that it interferes with deposition of material on the parts.

Manually operated immersion lines should allow for water spray rinses at three possible stages. In the case of water-washable penetrants a spray rinse must be available. Special agitated immersion rinses, where successful, qualify as semi-automatic lines. The second point where a water spray rinse is needed is at the pre-rinse stage of the post-removable penetrant process, and finally a spray rinse is needed at least as part of the final rinse after application of a hydrophilic or lipophilic remover. The use of immersion processes alone in any of these three instances can be expected to lead to difficulties in inspection and interpretation of results.

Application of penetrant remover on manual lines can be by spray and drain or by immersion and drainage. Immersion is by far the most common method of application and a tank will be needed for this.

Many manual lines provide a simple tank for the developer station. This is quite satisfactory if the developer is of the aqueous solution type. Dry powder developers are kept in such tanks but this cannot be regarded as entirely satisfactory as the developer in such an arrangement will get wet and become contaminated. A powder storm cabinet or a spray booth is a much better arrangement. If a non-aqueous suspension developer is used, a spray station for the developer is essential.

Just because nothing is applied to the parts in the drier station it does not follow that care is not necessary at this stage. In fact this is a critical stage in the penetrant process. Incomplete drying can be a nuisance, and incorrect drying can cause a total loss of effectiveness of the system. In the simplest lines a warm air blower can be used with great success. However, a drier station frees an operator to perform some other task. Still hot air is almost useless for the process. Warm (60–80 °C) mobile air recirculating over the surfaces of the parts is ideal. Care must be taken in the design of driers to ensure uniform heat and movement throughout the cabinet and to avoid hot or cold areas.

Inspection booths should be designed to allow operators to work in comfort. Where colour contrast penetrant testing is carried out on a line, a clearly defined area with good illumination (minimum 500 lux at the surfaces to be inspected), good ventilation and space for work to be organized in an orderly flowthrough must be provided.

When fluorescent indications are to be inspected white light must be excluded from the area as far as possible. Ventilation of closed inspection booths is very important in order to maintain operator comfort and efficiency. Such inspection booths must be equipped with UVA lamps (black light) which should give a minimum energy at the inspected surface of 1000 $\mu W/cm^2$ at a wavelength of 365 nm.

The remaining parts of a manual line are the rest and drain stations. These must be included in any penetrant line unless the number of parts to be processed per hour is very low and it does not matter if stations are occupied during cooling, drainage or contact time. It is normal to find these stations at the following locations:

1. after degreasing – for cooling;
2. after penetrant application to allow proper contact time;
3. after the final wash to allow as much of the water as possible to drain away before drying;
4. after application of developer (if aqueous wet developers are used this will be the same as point 3). Developers need time to work and a rest station before the inspection booth is very useful in avoiding the need to clutter the booth with waiting parts.

Waiting stations of these types must be large enough to take the load of parts and, where wet materials such as penetrant or water washings are draining off, must be equipped with a collecting tray shaped to run the drainings off through a drain hole for collection and disposal. It is also advisable to have a collecting tray under the dry waiting stations so that the area under the line is kept clean.

7.3 SEMI-AUTOMATIC PENETRANT LINES

Any one of the active stations on a penetrant line or the handling arrangements can be individually automated. The commonest form of automation is probably found in developer stations where there is an automatic cycle of dry powder storm application followed by removal of dust from the air inside the cabinet, thus allowing the lid or doors of the storm cabinet to be opened without a substantial escape of developer. This has several advantages. First, airborne dust is a nuisance and regular release of powder developer into the air around a penetrant station would make the area most unpleasant to work in. Second, developer is not wasted so readily. A further advantage with a recycle system is that active collection of the developer makes it easier to keep it dry and in acceptable condition.

After the developer station the most common form of added automation is found in handling the parts through a line. Overhead rails with hoists are very useful and widely used, despite the fact that such an approach means that any closed stations such as powder storms and driers must be split longitudinally to allow the hoist to pass through or must be of a top loading design with parts being released from the hoists at such stations. Overhead hoists are very widely used with spray booth systems where the problems of split roofs are minor. Another form of automated handling is the use of roller tracking which can be partially or wholly motorized. This is very commonly used on immersion lines where it is combined with pneumatic lift arrangements at the stations where parts are actually immersed. Hydraulics are best avoided since a leak of hydraulic oil could lead to the loss of several hundred litres of processing chemical.

It is also quite common to find wash stations including some form of automation. Spray heads arranged on stand-pipes are often very helpful at these stages of the process. It is usual on semi-automatic lines for such automatic washes not to be complemented by the use of a hand-held spray gun to finish off the process. It is particularly important in the case of water-washable penetrant systems to remember that water washing should last for only 2 min and this will include both the automatic cycle and any manual finishing off with a spray gun.

If any other stage in the process becomes automated it is usual to go all the way and automate the whole line right up to inspection.

7.4 AUTOMATIC PENETRANT LINES

The introduction of any automation in a penetrant line is usually made on economic grounds when large numbers of parts of similar size and shape are to be processed.

Full automation of processing normally infers very large numbers of parts. In some industries, notably the aerospace industry, the decision to automate all or part of a line is often taken with the view that greater repeatability of processing can be achieved and the limits of variation in processing parts can be reduced. There has been much work for many years on automation of inspection, thus releasing the inspector from his dark booth. Success was achieved some years ago and now even fairly complex shapes can be inspected automatically (Fig. 7.3). Much of this work is still experimental and the cost remains high. However, this approach has the attraction that images of critical parts can be recorded for later reference.

Fully automated lines may transfer parts on a conveyor either overhead or on roller tracks. Robotic manipulators in either linear or circular mode can also be use with success. A further possibility is to use an endless belt with parts travelling through and between stations at such a speed as to allow proper contact times. When parts or carriers with parts on them are processed automatically the load may be immersed or the tank of chemical lifted to cover the parts. Any of the approved methods for application of materials, either singly or in combination, can be used on automatic lines. In view of the level of control which can be exercised over the actual steps, it is possible to vary the actual contact or processing times from those laid down by the specifications for manual lines.

Fig. 7.3 A classic on-line automatic penetrant processing installation. Manual interference is minimal, with operators only required to load and unload carriers of components.

Automatic penetrant lines can be designed for single penetrant processes, several levels of sensitivity of the same process or even combinations of several processes.

The first major consideration when designing automatic penetrant lines is the presentation of parts. No amount of clever engineering can overcome the problems set by poor presentation of parts. Once this has been dealt with, the steps in the processes can be approached with some expectation of success. When penetrant is applied by immersion there is little chance that it will not reach surfaces other than areas where parts are touching fixtures or possibly each other, although penetrant can normally be expected to migrate between surfaces which are simply touching. When any spray-type application is used it is advisable to locate UVA lamps over the drain station so that a visual check can be made to ensure that penetrant has reached all surfaces. If a colour contrast penetrant is in use on such a penetrant line, such an inspection will be under white light.

Automatic water wash stations present problems of their own. Whether these stations are for complete removal of surface excess water washable penetrant or as pre- or post-rinse in a post-removable process, it is essential that an efficient spray rinse is achieved. This frequently involves close study and experiment with spray patterns and with rotation of parts or loads, and the use of spiral or other patterns of motion.

Remover stations also require close attention to achieve the desired results. This area is one where the whole process is most vulnerable. Where the processing leaves the charge of parts at stations it is advisable to ensure that a carrier remains with the charge during the entire contact time for the remover. Developer stations are readily automated. When dry powder developer is used application can be by way of a powder storm or electrostatic spray. Liquid developers of all types can be applied by spraying and in the case of aqueous developers these can be applied by immersion.

Automation of penetrant processes has been satisfactorily achieved in a number of ways for many years. Advantage must be taken of modern achievements in automation in order to obtain repeatable results economically.

7.5 DISPOSITION OF PENETRANT LINES

Floor space is at a premium in most factories and anyone given free choice of space in siting a penetrant line can be considered fortunate.

Penetrant lines can be set out in a straight line, an L-shape, a J-shape, a U-shape, an E-shape, an O-shape or a complete circle. In order to locate a line in a pre-existing space it may be necessary to use two lines of stations back to back. Manual lines are the most flexible and semi-automatic lines place little restriction on the disposition of stations. Fully automatic lines often require a return loop for empty carriers after the final stage, and simpler shapes are favoured so that complicated handling arrangements can be avoided.

An important factor is that however the various stations are disposed with respect to each other there must be access to all services such as compressed air supplies, water supplies, drain cocks, valves and electricity supplies. If a line is sited so close to a wall that access to some service at the rear of the line is difficult, periodic servicing of the line will be affected. While every effort can be made in construction of the line to ensure easy access, some services cannot be sited at the front.

7.6 REQUIREMENTS OF INDIVIDUAL STATIONS

Equipment should be so designed as to allow and encourage uniform, controlled and efficient operation of the penetrant process. Provision must also be made to allow easy access to all services and allow the line to be emptied, cleaned up and refilled from time to time.

7.6.1 Degreasers

Degreasers associated with penetrant lines are usually either vapour degreasers using a volatile none flammable solvent such as trichloroethylene or the alkaline type. Vapour degreasers are by far the most common. Parts are lowered into the degreaser tank in the base of which the solvent is boiling. The vapour from this solvent rises through cooling coils which cause the vapout to condence and then fall back into the liquid. If parts are lowered into the vapour phase the solvent vapour condenses over them, washing over the surface and taking oils, grease and some other organic materials with it.

When the parts reach the temperature of the vapour condensation stops and no further cleaning action takes place. A possible danger in this type of cleaning is the introduction of water which can lead to acidity in the solvent. Some authorities object to the use of halogenated hydrocarbons on titanium or stainless steels. Under such circumstances an alkaline degreasing process can be used. Parts are sprayed with or immersed in a weak (about 10 wt%) solution of inhibited alkali in water. This is then washed off with water and dried. When application of the solution is by immersion this becomes a three-station process. This can be reduced to two stations or even one if spray application is used. Great care is needed in treatment of effluent from alkaline degreasers.

With the current concerns over the use of volatile organic compounds other water based materials such as detergent solutions are becomming widely used. In common with the alkaline solutions their application must be followed by a water rinse and thorough drying.

7.6.2 Waiting stations and drain stations

Waiting and drain stations are present at various points on a penetrant line (Fig. 7.4) Waiting stations occur after degreasing, drying stations and developer applications. Construction is usually very simple, and is often just a section of roller tracking or grille fitted into a frame with a catch tray underneath. Drain stations are needed after penetrant application, after final water rinsing, whether the wash is from a water-washable penetrant or the final water rinse of a post-removable process, or after application of an aqueous developer. The construction of a drain station is normally similar to that of a waiting station except that the catch tray should be shaped to direct the liquids to a drain hole where they are collected for disposal.

Drainings from the penetrant contact area should not be directed back to the tank as their characteristics will have become altered by the time that they arrive there and they may introduce contamination into the chemicals.

Both waiting stations and drain stations should be designed so that the grills or rollers and the trays underneath can be readily removed for cleaning; they must also be robust enough to take any load likely to pass through the line. A load capacity of 100 kg is normally sufficient, but occasionally 350 kg loadings are required.

Frameworks can be of painted steel but drain trays, grilles and rollers are best made of stainless steel to avoid rusting. Aluminium trays should not be used with steel grilles or rollers as the combination of aluminium, water-based material and ferrous metal leads to severe corrosion problems.

Fig. 7.4 A typical 'tunnel' type station on a penetrant processing line. In this example a basket of turbine blades is waiting to be transferred to a drying oven. Note that the basket has individual places for each blade. This arrangement allows uniform processing and avoids the problem of 'touch marks' which will occur if the components are simply loaded into an open basket.

7.6.3 Penetrant stations

The actual design of the penetrant station depends on the method chosen for application. Penetrant applied by immersion will involve a tank whether the line is manual or automatic. Penetrant tanks can be made of a wide variety of materials. Penetrants are non-corrosive and so mild steel can be used. Polypropylene has been used but plastics are not a good choice for penetrants as many of them contain plasticizers which, in extreme cases, can dissolve the tank with inconvenient results. Although penetrants do not corrode the tanks, the air around penetrant lines is often very humid after water spraying and parts of the tank exposed to the air will rust.

The best material is stainless steel. Penetrant tanks should be equipped with two drain cocks, one in the base and the other about 10 – 15 cm from the base depending on the size of the tank. This is to allow the clean re-usable penetrant in the upper part of the tank to be drained off through the upper outlet and retained while the sludge in the base of the tank can be taken out separately and disposed of. Ideally, penetrant tanks should be fitted with lids to avoid contamination from overspray at wash stations or other sources.

Fig. 7.5 Electrostatic application of fluorescent penetrant to an aircraft wheel. The wheel shape is very suitable for the use of electrostatic spray application of both penetrant and developer.

When penetrant is pumped through a nozzle for a flow-on technique or is sprayed on, either conventionally or electrostatically, the penetrant station takes the form of a booth on a manual line or a cabinet on an automatic line (Fig. 7.5). In the case of electrostatic spray there must be 600 mm clearance between the maximum dimension of the parts and the spray booth or cabinet walls. When any form of spray application is used there must be some form of extraction even in automatic stations. In manual operation the need for efficient extraction is very clear. Water curtains work but are rarely used in penetrant spray booths. Extraction is usually by air movement. The most effective form draws air from the operator across the parts to be sprayed, taking overspray out through the back wall. Such systems require an extraction with a face velocity of around 0.7 m/s. The filters through which extraction is carried out should be removable for cleaning or exchange.

Spray booths must allow access for handling the parts. The roof may be in two parts leaving space for access from an overhead crane. Spray booths must be constructed of corrosion-resistant material such as stainless steel or a good grade of galvanized steel. The interior design and the extraction must be designed in such a way that swirling of the sprayed penetrant is avoided. When this phenomenon occurs it can be very difficult to cure and leads to penetrant-covered operators as well as or instead of penetrant-covered parts.

By its very nature fog application of penetrants is restricted to automatic lines. There are enough ways of covering operators with penetrant without attempting this process manually.

7.6.4 Water wash stations

Water wash stations may occur at three points on the penetrant processes. All three need great care.

In the water-washable penetrant process the water wash must normally include a spray. Only in very rare circumstances is immersion alone, agitated or not, sufficiently effective, and the water is fouled with penetrant very quickly so that control of the process is lost. Spraying with fresh water each time is by far the best method of water washing at this stage. An immersion tank with a weir can be useful to help with this process. The spray tank should be deep enough to prevent splashing, and splash guards should be provided to avoid overspray to other stations on the line.

Air and water hoses to hand-held guns should be oil resistant. The spray gun should be of a design that allows easy one-hand operation. It must be possible to arrange the spray pattern to allow good delivery of moderately heavy droplets at a pressure regulated below 2.8 bar (40 p.s.i.). The arrangements for automatic spray washes must be made very carefully to ensure a good wash. Components or charges on a carrier can be rotated or moved in a spiral within the spray to assist the process.

In all washes the water temperature should be kept below 35 °C, and ideally at 25 ± 5 °C. In the case of water-washable fluorescent penetrant lines a low energy UVA illumination (300 $\mu W/cm^2$) must be provided to allow assessment to be made of the effectiveness of the wash.

In some wash stations the effluent is taken to some kind of filter (section 5.3.4) and returned to the spray gun for re-use. This is a dangerous practice and should be avoided. Water-washable penetrants contain an emulsifying agent and these penetrant washings are dilute solutions of emulsifier. Should the filtration cease to work for any reason the water will return to the spray station untreated and the concentration of emulsifier will increase with each cycle, thus washing off the penetrant with a solution of emulsifying agent. The least that can be said of this is that the situation is out of control.

Water spray rinses are used on post-removable penetrant lines as a pre-rinse to remove the majority of penetrant before use of the remover solution. This stage *must* be a spray rinse, whether it is manual or automatic. Simple or agitated immersion does not work. Conditions for the spray rinse in this instance are the same as for those for water-washable penetrant systems. The third instance of a water rinse is the stage after application of a remover solution or emulsifier. In these cases the ideal water wash involves immersion in a bath containing enough water to stop the remover or emulsifier action followed by a spray.

Wash stations are best constructed of stainless steel, thus avoiding corrosion problems. Aluminium grilles must be avoided as they will start to corrode when in contact with water and ferrous metal. This is not to suggest that aluminium parts cannot be processed on a penetrant line – they can. However, it is best that they are carried on aluminium fixtures. The contact time of the parts with water is generally too short for electrochemical action to be serious, and the parts are actively dried. However, grilles are left in the station and are not dried, and if they are made of aluminium they will corrode.

7.6.5 Remover stations

Currently great interest is being shown in spray application of hydrophilic remover solutions even to the extent of using electrostatic spraying. Most application is by immersion with or without agitation. Most stations are simple tanks which should be made of stainless steel to avoid corrosion and equipped with a drain point.

If agitation is allowed or required, it is best provided mechanically. Air could be used, but it is chemically active and gives weakly acidic solutions, thus introducing a further element of lack of control. Spraying solutions of detergents leads to foaming which can create disposal problems. In the case of lipophilic emulsifiers, a simple tank can be used and, since these materials contain no water, mild steel can be used in construction.

Aluminium grilles will corrode in a water-based solution of hydrophilic remover and so must be avoided. In the case of emulsifiers this risk is much less but it is still a good principle to avoid electrochemically active mixtures of metals.

7.6.6 Driers

Driers are the most neglected stations in penetrant processing lines. Mobile filtered hot air should be circulated over the wet surfaces in a way which ensures rapid efficient drying so that the surface temperature does not exceed 60 – 70 °C on manual lines. The advent of heat-fade-resistant penetrants means that the possibility of drying at higher temperatures and thus in shorter times can be investigated. Temperatures of about 100 °C at the surface are still a practical maximum.

Driers must be constructed in such a way as to avoid heat losses through the walls. On small lines a cabinet with a curtain can be used. On larger lines a tunnel type drier with rising insulated doors with seals can be used. Top loading driers with sealing lids also work well. Hot and cold spots must be avoided in the design of driers.

7.6.7 Developer stations

The design of developer stations varies with the type of developer in use. Dry powder developers are normally applied by powder storm or spray gun. Only when very small numbers of components are processes can an open tank be justified. Powder storms should be equipped with a heating arrangement to reduce humidity in the cabinet and conserve the developer. Dry compressed air or mechanical agitation can be used to disperse the powder, and an extract system is needed to remove airborne powder from the cabinet after a good storm has been produced to allow the station to be opened without the nuisance of a dust cloud and loss of developer. Compressed air agitation can be used to blow back such extracted developer powder. Construction can be of the tunnel form or a top loader with lid. Dry powder developer can also be applied using a flock gun or a fully electrostatic spray gun (Fig. 7.6). Application to large components in a spray system is best achieved this way.

Non-aqueous suspension developers must be applied by spray and, except where aerosols are used, this requires a spray gun with a cup or reservoir which will agitate the suspension to a uniform consistency before spraying. In installations of this type good extraction is essential to protect operators from solvent fumes.

Fig. 7.6 This large ring-shaped component is an ideal subject for processing using electrostatic application apparatus. In this case dry powder developer is being applied manually. Note the earth lead and the operator's gloves.

Developers which are solutions in water can be applied by simple immersion in which case a simple tank with a grille and drain point is needed. This type of developer can be applied as a wet spray. If the spray is too fine the solution will dry out on its way to the part. Such application can be carried out in a spray booth.

Suspensions of developers in water are usually applied by immersion with agitation. This requires a tank with an agitation system. Mechanical agitation is preferred since air is chemically active. The drain point on such tanks must be large enough to allow both the solid and liquid contents to be thoroughly cleaned out. Water suspensions can be applied by spray. Despite the fact that all water-borne developers contain corrosion inhibitors, all apparatus and material associated with these stations must be corrosion resistant. Stainless steel is preferred.

7.6.8 Inspection booths

The nature of inspection booths will vary depending on whether a colour contrast penetrant system or a fluorescent system is in use. Both types require an area which is large enough to allow controlled inspection of processed parts. This will entail a well-ventilated comfortable area with adequate table space.

The inspection booth for a colour contrast line requires good white light illumination – a minimum of 500 lux at the part surface. Low power magnifiers (5× or 7×) should be available.

In a fluorescent penetrant line as much white light as possible should be excluded. Some specifications require ambient white light to be as low as 5 lux. The imspection booth on the smallest lines can be a table-top version. On medium sized lines a curtained booth is used succesfully, whereas in very large penetrant departments where there are several lines a complete darkroom can be set aside for inspection. Individual inspection stations must be equipped with UVA lamps (black light) which give a minimum illumination of 1000 $\mu W/cm^2$ at the part surface.

The dark booth must also be equipped with white light to allow a final inspection for defects so large that no penetrant is retained in them and for inspection of indications under low power magnification (5× – 10×).

At present there is considerable interest in automatic inspection. The attractions are that the images obtained can be processed and stored, that the machine does not tire and that once set up inspection will be consistent. Automatic inspection has been achieved in two ways – by conventional optical scan and by laser scan. In both cases the image can be stored readily and also displayed on a screen. Both systems are costly.

7.6.9 Fixtures and accessories

(a) Baskets and jigs

Except when large parts are processed on a hook it is necessary to use baskets and jigs to carry parts through a penetrant line. The construction of these must be strong enough to carry a normal load. They should not be overloaded as this will cause poor processing as well as strain the equipment. The choice of material is important. Since water is involved in most penetrant processes, care must be taken to use carriers which will not react with the parts being processed in water. Contact with jigs is so close that, for critical parts, even the short contact time with water and active drying after the wash cannot guarantee that there will be no reaction. Electrochemically similar metals for jigs or inert fixtures coated with polytetrafluoroethylene (PTFE) will overcome the problem.

(b) Timers

The contact times of each stage of the penetrant process must be controlled, and they are critical in the penetrant removal stages and the drier. Timers with large sweep dials for ease of reading must be supplied at each station. They should provide an audible as well as a visual signal. On automatic lines it is normal to use indicating lamps and a visual and audible warning of low levels of chemicals coupled with a stop should any station fail to function.

(c) Air lines

Low pressure air lines of 0.35 bar (5 p.s.i.) can be used between water washing and drying to remove water droplets and push water from blind holes, through holes, threads, and keyways, thus assisting the drying process. Such air lines can also be used to remove excess dry powder developer before inspection. The air must be clean and dry, and air lines must have traps to remove contamination.

(d) Spectacles

Some inspectors find that sodium glass spectacles help in inspection and give them better contrast between the indication and background. If used they should be optically neutral or of the lens type normally worn by the inspector if he uses spectacles. Photochromic spectacles should not be exposed to UVA (black light) since this can cause irreversible tinting of some types.

7.6.10 Effluent control

The effluent from penetrant lines is not particularly harmful but it is coloured and therefore readily recognized. The major source of effluent is contaminated water. This comes from three sources:

1. washings from water-washable penetrant systems;

2. washings from a pre-rinse station on a line using the hydrophilic post-removal system;
3. washings from post-rinse stations on either a line using the hydrophilic post-removal system or a line using the lipophilic post-emulsifiable system.

The washings from 1 and 3 contain oil-based material – penetrant and/or lipophilic emulsifier chemically emulsified in water. In the case of the hydrophilic remover system post-rinse they also contain detergent – in fact they are mainly detergent. Such a mixture cannot be seperated physically and must be chemically, a flocculant process or absorbed on to active carbon or dealt with actively in some other way. Therefore carbon filtration is generally used. There are many forms of active carbon designed for different purposes and care is needed in selecting an effective type. The washings from 2 – the pre-rinse of the hydrophilic system – present a different problem. If the penetrant is truly hydrophobic, it is possible to clean the water with a coalescing filter. If the penetrant has some water tolerance, the coalescing filter will only be partially succesful and must be followed by a further treatment. Even when penetrants are genuinely hydrophobic a further treatment polish is used to remove the final traces of penetrant as contaminants in the system can cause a small degree of chemical emulsification of the penetrant. (This topic is discussed more fully in Chapter 11.)

7.6.11 Optional Hot Dip

On some penetrant lines a hot dip in water (80–90 °C) is included before drying. This is to aid drying. A stainless steel tank with thermostatically controlled heating is needed for this stage.

QUESTIONS

1. When designing a penetrant installation it is necessary to ensure that the process does not hold up the flow of components unnecessarily. To achieve this
   (a) the number of components per year and the number of working hours in the year must be known so that the throughput can be calculated
   (b) the need for jigs and baskets must be calculated and 20% extra allowed for unforeseen demand and damage
   (c) the maximum throughput at any one period of time must be known and that figure used to design the throughput
   (d) the possibility of automating one or more of the stages must be considered

2. Field inspection kits generally contain a colour contrast penetrant process. This is because

(a) obtaining the levels of darkness needed for inspection of fluorescent indications is often very difficult in field conditions
(b) fluorescent penetrants cannot be used satisfactorily in aerosols
(c) it is difficult to obtain electricity in the field to provide power for a UVA lamp (black light)
(d) operators with less training and experience normally carry out penetrant inspection in the field

3. In order to ensure a long trouble-free life, the wet stages of a penetrant installation should be made from
   (a) stainless steel
   (b) glass-reinforced plastic
   (c) aluminium
   (d) carbon steel

4. Grilles in the base of wet station tanks (particularly where water is involved) and jigs or baskets which are in very close contact with critical components must be made from an inert material or a similar metal to the tank in the one case and from the component in the other. The reason for this is
   (a) mixed metals make the penetrant become corrosive
   (b) mixed metals in a water-based environment cause an electrochemical couple to form and corrosion of one of the metals
   (c) mixed metals may suffer corrosion when in contact with dry powder developers
   (d) it does not matter if mixed metals are used

5. Dry powder developer stations on installations designed for a large throughput may use
   (a) a powder storm
   (b) an electrostatic spray
   (c) either of these methods
   (d) neither of these methods

Note: There is only *one* correct answer to questions 1–5. Tick your choice and check it with the answers on p. 215

8
Quality control of penetrant lines and control checks

Penetrant testing is itself a form of quality control. It is very important that materials and equipment for the process are subjected to strict quality control from the first stages of design through field testing, manufacture, supply and use right up to the time that such materials and equipment are scrapped. The quality control measures required during design are not dealt with here. The quality control measures used in manufacture are discussed here because this is not simply of general interest but relates directly to the tests and controls applied during the actual use of the materials and equipment.

All serious manufacturers of any item have an internal quality control programme which relates to some national standard such as British Standard BS 5750 in the United Kingdom, ISO 9000.

8.1 QUALITY CONTROL OF PENETRANT CHEMICALS DURING MANUFACTURE

Penetrant chemicals can be divided into three types:

1. penetrants themselves
2. developers
3. penetrant removers

These can be further subdivided into specific types.

The tests performed during manufacture fall into four types:

1. functional tests
2. material conformity tests
3. safety
4. economic factors

8.1.1 Functional tests

Functional tests are carried out either on the whole penetrant system or on a single part of it. The keys to functional tests are good test pieces or (better still) parts with known defects in them and materials of known performance.

The first test will usually be of the entire system in sequense using the newly manufactured material and the used materials, if the results are equivalent, the material is satisfactory for that particular test. It is worth noting here that the choise of test piece for this type of test is critical. If a test piece has cracks which can be found using a low sensitivity penetrant testing, it is no use testing higher sensitivity systems on such pieces as they will all look the same.

At present a number of test pieces or sets of test pieces are available which are very valuable for this work (Figs 8.1–8.5). The TAM panels with five star-shaped crack patterns require increasing sensitivity from the system are an example. There are two sets available which consist of twinned pairs of test panels with certificated crack depth, and these allow sensible application of functional tests.

Care of test panels is extremely important. Penetrant residues must be

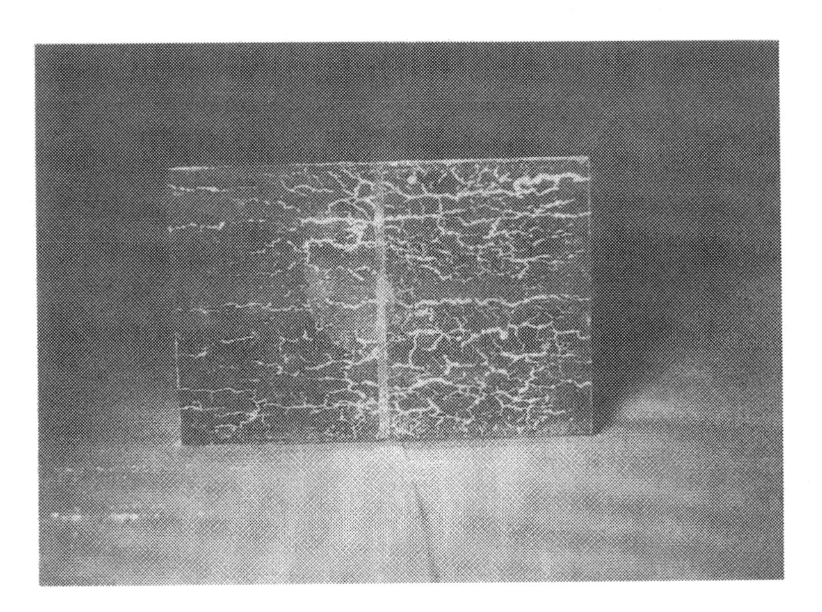

Fig 8.1 An artificially cracked aluminium test panel. This type of test panel is now less frequently used as the cracks are easily found by use of quite moderately sensitive processes. In this instance a standard sensitivity fluorescent penetrant has been applied to the left-hand side and a much diluted one to the right-hand side. Despite the slot in the middle the more sensitive material has migrated to the right-hand side during the penetrant contact time and it is only in the right-hand quarter of the panel that the potential of the diluted penetrant is seen.

Fig 8.2 Simple aluminium test panels for checking penetrant process performance

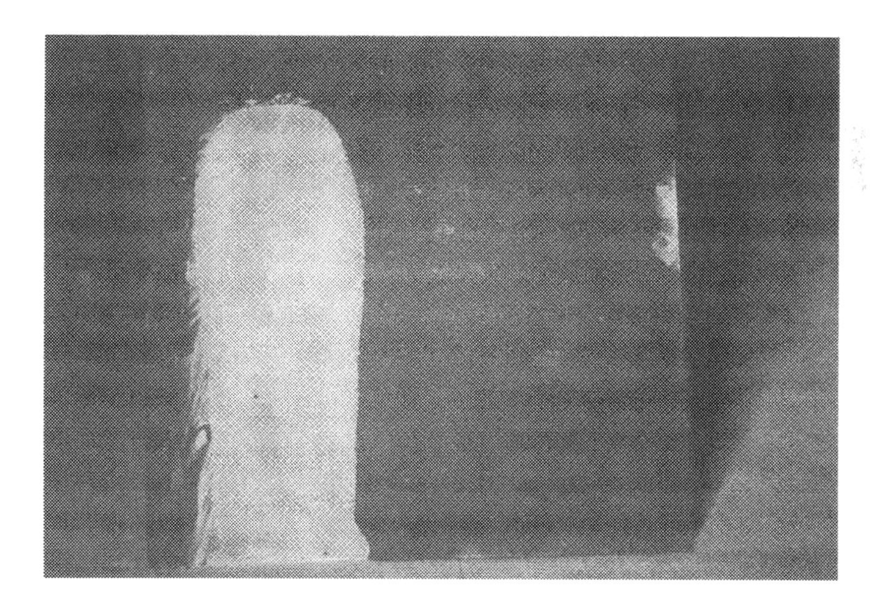

Fig. 8.3 The wash response of used penetrant can be tested on a sandblasted stainless steel panel. The sample on the right failed.

Fig. 8.4 This stainless steel test panel is an excellent tool for day-to-day checking of the penetrant process. On the right-hand side five star-shaped indications have been manufactured in a smooth satin finish area while the left-hand side is grit blasted to allow the level of penetrant removal to be checked.

cleaned out thoroughly directly after each use, and the effectiveness of each clean-up must be checked by applying developer and waiting for 30 min to see whether any indication forms. Failure to do this will lead to a situation where the indication from a previous use will appear and this is not really of much use. Eventually, the cracks will clog up so much that only a halo will appear where there should be an indication. If any indication is found after clean-up the test piece must be recleaned. The standard cleaning method is to apply a solvent spray or to wash the test panel and physically remove developer residues. If this is insufficient, boiling for 30 min in a 50% solution of detergent followed by rinsing in demineralized water and then in acetone usually works well. Sometimes ultrasonic cleaning is recommended. However, since penetrant residues are rarely solid they absorb the energy and results are uniformly disappointing.

Test panels are costly and care is needed in storage. Storage in dry acetone is best.

Penetrant removal and tests for the efficiency of penetrant removers involve the use of test pieces which are essentially sand-blasted stainless steel plates. Sush tests involve comparison with standard materials and require standardized procedures to be useful.

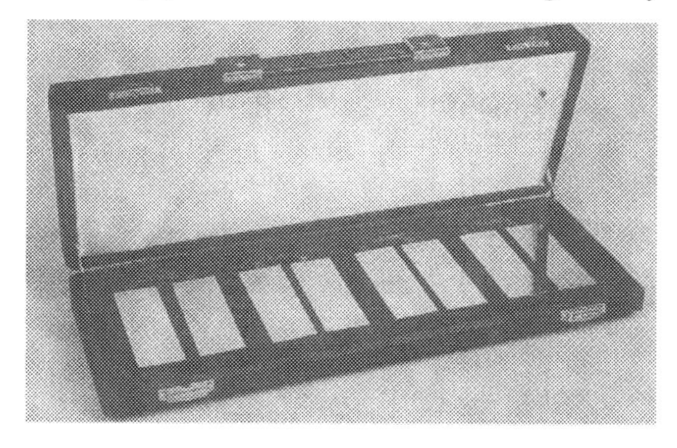

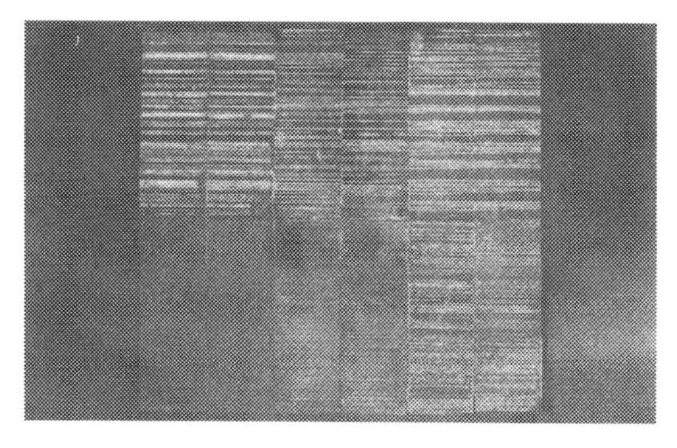

Fig. 8.5 Modern test panels with certificated crack depths and measurable crack widths are very useful for comparison of processes and materials. These nickel-chrome plated test panels are manufactured in twinned pairs and allow very useful laboratory work in the development of materials: (a) in their case; (b) after processing.

8.1.2 Material conformity tests

Material conformity tests are designed to ensure batch-to-batch conformity in the production of penetrant chemicals. No single test is definitive, but a programme for each material including up to as many as 15 different tests provides the assurance that materials of the same name are essentially the same. Such tests include the flash point of flammable materials, viscosity, specific gravity, fluorescent brilliance, colour intensity, particle size of developers and other factors. Tests are also carried out for fluorescent fading and heat fading of fluorescent penetrants.

It is difficult to classify tests for residual sulphur, halogens and alkali metals. Some authorities would consider them to be safety tests, and this classification is probably justified in the nuclear industry. In other industries the justification is less clear. There are a number of methods for measuring sulphur and halogen levels which are fairly accurate down to 100 ppm. Specifically, the alkali metals can be measured down to a few parts per million.

It is obvious that finished products cannot be expected to be within specification unless the raw materials and intermediates are similarly controlled. In manufacturing, a full programme of testing which controls raw materials from reception to requisitioning, total manufacture and eventual supply to the user must be instituted and audited.

8.1.3 Safety tests

Safety tests fall into two categories: operator safety and safety of the components being tested. Operators should not be exposed to harmful chemicals. The only penetrant chemicals which cause real concern are the volatile solvents used in the solvent removal penetrant as removers or carriers for non-aqueous suspension developers. These are harmful if inhaled, flammable with flash points around 10 °C or are both harmful and flammable. Careful application of these materials in a well-ventilated area or with respiratory protection allows safe usage.

The safety of personnel is considered during formulation and materials of known or suspected toxicity are avoided. A continuous supply of new toxicological data on chemicals is available, and any manufacturer of chemicals should monitor this closely. Aerosol packages have their own hazards as they are under positive pressure. Storage of aerosol cans at high temperatures, i.e. above 50 °C, can cause problems.

The safety of the parts under test has been considered from the earliest days of penetrant testing. Some early emulsifiers produced quite alkaline mixtures with water. This is avoided in modern materials: both the detergent removers and the modern lipophilic removers give neutral solutions in water.

Aqueous developers are formulated to avoid corrosion problems. Corrosion tests are carried out on all penetrant chemicals using a wide range of materials such as carbon steels, aluminium, magnesium, nickel-based alloys, cobalt-based alloys, stainless steels, titanium and other metals to ensure that penetrant testing remains non-destructive.

8.1.4 Economic factors

Economic factors in penetrant testing reach far beyoond the simple calculation of raw material costs, although these can be a serious consideration. Stability in open tanks during normal use must be considered. Stability to temperature change is important as some mixtures separate into several layers at low temperatures. Other mixtures are sensitive to heat and some chemicals evaporate even at quite moderate temperatures.

Contamination is the greatest threat to penetrant materials. The usual contamination in penetrant stations is water, although cleaning solutions and organic solvents also find their way into the penetrant from time to time. Of these contaminants cleaning solutions containing chromates, nitric acid or ferric chloride are by far the most objectionable; they destroy fluorescence when present at quite low concentration. Solid rubbish such as metal dust, cigarette ends, plastic coffee cups and other casual contamination find their way into penetrant tanks and emulsifier tanks without causing many problems.

A major contaminant in remover stations is, of course, penetrant. The systems are designed to deal with this and the remover bath must be changed occasionally as part of the management of the process.

Developers are designed to resist contamination. Water is the curse of dry powder developers, and heaters inside developer stations help to conserve the state of the powder.

8.2 PERIODIC CONTROL OF PENETRANT CHEMICALS IN USE

In this section we discuss procedures needed to ensure that used materials are still fit for use. In installations where the penetrant chemicals are used only once, e.g. spray lines, or when aerosols are applied, such tests are superfluous unless you do not trust your manufacturers.

8.2.1 On site

In this context 'on site' means on the line, wherever it might be. Many tests are fairly simple and can be carried out on a bench by the line or using the line itself. Other tests require laboratory apparatus and personnel skilled in its use. This second group is discussed in section 8.2.2. The tests which can be carried out on site are described below and are listed in Table 8.1.

Table 8.1 Frequency of on-site control checks on penetrant chemicals and acceptance limits

Test	*Frequency*	*Limits*
Functional test	Daily	Equal to standard
Fluorescent brightness	Weekly	Equal to standard
Colour	Weekly	Equal to standard
Concentration of hydrophilic remover	Weekly	±1% of nominal concentration
Remover/emulsifying agent efficiency	Monthly	Equal to standard
Penetrant content of emulsifying agent	Quarterly	10% contamination may cause problems
Fluorescence in developers	Weekly	None
Concentration of aqueous developers	Weekly	±3% of nominal concentration
Coverage by aqueous developers	Weekly	Equal to standard

(a) Functional tests

A functional test should be carried out each time the line is used by including a test piece with the first load. Comparison with a photographic or transfer lacquer record of the test piece will show whether the process is working. A more detailed functional comparison can be made by comparing the used materials with retained samples. When a line is filled or any of the stations are replenished or the contents are replaced, a sample of the fresh material must be taken into a new screw cap tin. Polythene bottles can be used for removers or developers, but not for penetrant which should be stored in tins or glass bottles. If the bottle is of clear glass the sample should be stored in the dark.

All retained samples should be labelled clearly showing the identity of the product, the manufacturer, the manufacturer's batch number and the date of fill. When taking samples of used materials it is necessary to ensure that they are representative of the tank.

The functional test is very helpful as it relates so closely to the actual inspection process. If the overall process is found defective, it is usually fairly simple to isolate the cause of the problem by substituting standard material at each stage whilst retaining the materials normally used for all other steps. When the standard materials give the expected results it is the substituted material which is suspect and needs attention.

Functional tests can be carried out on actual parts of the type or types normally inspected at the penetrant operation or on test pieces with artificially induced cracks or on both. Actual parts should have the smallest allowable defect in a known location. Artificially cracked parts obtained commercially or from a control laboratory are not readily reproducible from panel to panel and it is thus necessary to test half the panel using the materials under test and the other half using standard materials. The choice of test panel is very important as it must be of a type suitable for the materials used. Cracked aluminium blocks are very unlikely to show great variation with post-removable penetrant systems until the materials have become so grossly degraded for this to be visually obvious without the trouble of actual testing. However, highly polished chrome test panels are of little help in assessing most water-washable penetrants. This factor emphasizes the desirability of using both test panels and actual parts with known defects. Care must also be taken in the storage of test parts, whatever their type.

It is of little use to conduct tests on panels which already have fluorescent indications. Simple inspection under a black light will show whether the part is clean or not. Rigorous cleaning after use of test pieces of any type is necessary to ensure that they do not become filled with immovable residues – fluorescent or otherwise – and so useless. Ideally test parts should be stored under a solvent, but where this is inconvenient they can be kept dry in a clean place that is well covered and protected.

The nature of the functional tests is such that both processing and evaluation are critical, and so both the operator and inspector – ideally two different people – should be very familiar with both the process and the test parts.

(i) Functional test procedures

Post-removable penetrant system

1. Cover half the test panel with used penetrant and the other half with standard material.
2. Allow 10 min contact time.
3. Wash the bulk of the surface excess off in the normal way (immersion and/or spray) using water at a temperature between 10 and 40 °C.
4. Immerse in the concentration of hydrophilic remover used on the installation for not more than 2 min.
5. Wash with water (spray and/or immerse) at a temperature between 15 and 40 min.
6. Dry in a hot air drier until the piece is just dry.
7. Develop in the normal way. Note that if an aqueous developer is used, application of developer will precede drying.
8. Inspect the panel under black light after the normal development time.
9. Note any difference between the two halves. Where such a difference exists the fluorescence test (see below) will indicate its extent.

Post-emulsified penetrant systems

1. Cover half the panel with the used penetrant and the other half with standard material.
2. Allow 10 min contact time.
3. Apply standard lipophilic emulsifying agent to the panel for the specified time.
4. Wash with water at the temperature used on the penetrant inspection line (10–40 °C).
5. Dry in a hot air drier until the piece is just dry.
6. Develop in the normal way. Note that if an aqueous developer is used, application of developer will precede drying.
7. Inspect the panel under black light after the normal development time.
8. Note any differences between the two halves. Where such differences exist the fluorescence test (see below) will indicate the extent of this difference.
9. A further check is to clean the panel rigorously and then repeat the test using an extended emulsifier contact time (30 s more for a rapid emulsifier or 60 s more for a slow emulsifier). In each case fewer and fainter indications will be seen, but the relative appearance of the two halves should be the same.

Hydrophilic remover solution

1. Repeat the test for the post-removal penetrant system using test panels which have been covered with standard penetrant.

2. After the first water wash dip half the test panel in a fresh solution of remover at the correct concentration and the other half in the bath of remover being used. Ensure that the contact time with the remover is the same for both halves.
3. After completion of the process inspect the pieces under black light. If there is significant excess fluorescent background with the used material, replace the bath.

Emulsifiers

1. Repeat the test for the post-removal penetrant system using test panels which have been covered with standard penetrant.
2. Apply the emulsifier used to half the panel (vertical division) and unused standard material to the other half. Take care to use the correct dip and drain time for both halves.
3. After completion of the process inspect the pieces under black light. If the process is seriously impaired replace the emulsifier.

Water-washable penetrant systems

1. Cover half a test panel with used penetrant and the other half with standard material.
2. Allow 10 min contact time.
3. Wash with water at the same temperature as that used in the penetrant testing installation in the normal manner taking not more than 2 min.
4. Dry in a hot air drier until just dry.
5. Develop in the normal way. Note that if an aqueous developer is used, application of developer will precede drying.
6. Inspect the panel under black light and note any variation between the two halves.
7. Clean the panel rigorously and repeat the test using a 3 min wash time. This should cause overwashing and loss in both number and brightness of indications. However, the relative difference between the halves should remain constant.

Developers
Developers can be checked by processing test pieces with standard materials and then applying new developer to one half and used materials to the other. This procedure should be carried out irrespective of the type of developer used. Dry powders, while retaining their superficial appearance, lose sensitivity when used in a powder storm. This is because they are balanced mixtures of particles of varying size and shape, and many powder storms cause a disproportionate loss of the smaller particles.

(ii) Fluorescent penetrant brightness test There are two types of tests for fluorescent brightness. One is simple and can be carried out on the penetrant line, whereas the other requires laboratory instruments which is described in section 8.2.2. The on-site test requires two stoppered measuring cylinders, filter paper, tweezers and general laboratory standard methylene chloride.

The test procedure is as follows.

1. Put 1.0 ml of used penetrant into a measuring cylinder and fill to 100 ml with methylene chloride.
2. Stopper the cylinder and invert several times to mix the contents. Do not shake vigorously.
3. Repeat steps 1 and 2 using retained sample of the penetrant. Label the cylinders.
4. Cut a Whatman No. 3 or similar quality filter paper into squares of about 3 cm a side.
5. Mark the filter papers using a pencil. Then, using tweezers, immerse one in the solution of used penetrant and the other in the solution of standard material.
6. Allow the filter papers to dry.
7. Compare the fluorescent brightness of the two filter papers under a UVA lamp (black light) of energy 1500 $\mu W/cm^2$ at the filter papers. If there is a visually detectable difference there is cause for concern since the human eye can detect differences in fluorescent brilliance of more than 15%.

After this test the measuring cylinders should be washed out using laboratory grade acetone followed by a water rinse and a further acetone rinse.

(iii) Concentration of hydrophilic remover Water is quite volatile and in an open tank, a concentration of 5% remover can become 6% or even 7%, given time. An accurate check on concentration should be carried out every 6 months in the laboratory using a co-distillation method. On the line a rapid check can be made using a refractometer (Fig. 8.6). The refractive indices of low concentrations of remover are close to 1.33. However, a reliable calibration curve can be drawn between 3% and 10% provided that the refractometer is not colour sensitive. Calibration curves for new and contaminated remover should be prepared in the laboratory and then issued to the line.

The test procedure is as follows.

1. Dip a plastic rod into the remover solution. Avoid using glass or metal rods as these may scratch the glass surfaces of the refractometer. Note that the solution must be well mixed for accurate results.
2. Raise the cover plate of the refractometer and place two or three drops of solution on the prism face. Close the cover plate and ensure that the solution covers the prism face.

Fig. 8.6 Checking the concentration of hydrophilic remover by using a refractometer.

3. Hold the refractometer close to a bright white light source so that light illuminates the prism.
4. Look through the eyepiece and read the refractive index of the solution.
5. Read off the concentration of the remover from the calibration curve.
6. Wash the prism face and cover plate of the refractometer and wipe with a clean lint-free cloth.

(iv) Efficacy of removers and emulsifying agents The basic requirement of removers and emulsifiers is to remove surface excess penetrant while leaving penetrant in surface defects. Testing is carried out using standard penetrant material and standard remover solution or emulsifier on the grit-blasted stainless steel panels described in section 8.2.2 (a).

The test procedures are as follows.

Post-removable system

1. Cover the grit-blasted surface of two of the removability test panels with standard penetrant. Allow normal contact time for the installation.
2. Rinse with water in the normal way.
3. Immerse one panel in the remover solution bath and the other in a fresh

solution of the concentration usually used at installation. Allow 40 s immersion time.

4. Rinse the panels under ambient UVA (black light) and note any difference in the appearance of the plaques.
5. Replace remover solutions which no longer give satisfactory results.

Post-emulsifiable systems

1. Cover the grit blasted surface of two of the removability test panels with standard penetrant. Allow normal contact time used in the installation.
2. Dip one panel in used emulsifying agent and the other in standard material and allow to drain for the standard time used on the installation.
3. Wash with water under ambient black light noting any differences.

(v) Penetrant content of emulsifying agents Since it is rare to find emulsifying agents used with a water pre-rinse the process involves unavoidable carry over of penetrant into the emulsifier. By design emulsifying agents and penetrants are miscible in all proportions, although the efficiency of emulsifying agents falls when the level of penetrant contamination becomes high. Since emulsifying agents are normally more viscous than penetrants, drag-out tends to exceed that of penetrant, and the replenishment with fresh emulsifying agent which is then required imposes a natural control on the level of contamination of emulsifying agent by penetrant.

The test procedure is as follows.

1. Place standard emulsifying agent in a test tube.
2. Place used emulsifying agent in a second test tube.
3. Prepare mixtures of the standard emulsifying agent with 5, 10, 15, 20 and 25 vol.% of the penetrant under test and place similar volumes in similar test tubes to those used in 1 and 2.
4. View the specimens under UVA (black light) in a darkened area and match the colour.
5. Assessment of the usefulness of a contaminated emulsifying agent will be a matter of experience as the level of contamination which can be tolerated varies from penetrant to penetrant.

(vi) Fluorescence in developers

Dry developers

Any type of developer which is re-used may become fluorescent. Dry developers may be contaminated if they are exposed to parts which are not completely dry. This will lead to specks of fluorescent powder in the developer which may become deposited on parts and give spurious indications or simply make inspection very difficult.

The test procedure is to expose the powder to UVA (black light) in a dark environment. If fluorescent is detectable the powder should be replaced.

Aqueous developers

Since aqueous developers, in suspension or solution, are not protected by a drying stage they are more likely to become contaminated with fluorescent material. Fluorescence in these materials will reduce the contrast between indications and the background which can cause loss of sensitivity to small defects.

The test procedure is to dip one part which has not been processed in the used developer, repeat this with a second part in fresh developer, dry both parts and then inspect them side by side under UVA (black light).

(vii) Concentration of liquid developers The developing action of liquid developers depends on the nature and thickness of the layer formed. An important factor in the formation of this layer is the concentration of the developer.

Non-aqueous suspensions should not present any concentration problems as these materials should only be agitated thoroughly and then applied by spraying. Re-use by dipping should not occur.

Aqueous solution should change concentration only because of loss of water by evaporation. However, aqueous suspensions suffer not only this loss but also the disproportionate loss of solid owing to greater drag-out of solid during use compared with that for water and soluble constituents. The test procedure for aqueous solutions is to check the specific gravity using a hydrometer. This test can also be applied to aqueous suspensions provided that the suspension is not too thick to prevent proper operation of the hydrometer.

Hydrometer method

1. Mix developer thoroughly in the tank.
2. Place a sample in a glass cylinder.
3. Place a hydrometer of suitable range in the cylinder and allow it to come to rest without touching the walls.
4. Read the scale at the bottom of the meniscus.
5. Check the specific gravity according to the data supplied by the penetrant material manufacturer.

(viii) Concentration of aqueous suspension developer

1. Weigh a dry flat-bottomed container of capacity 200– 300 ml accurately to 0.1 g.
2. Add approximately 100 g (accurately known to 0.1 g) of well-mixed developer.
3. Evaporate to dryness on a hotplate or other suitable apparatus.
4. Remove from heat, allow to cool, reweigh and calculate the percentage of solids:

$$\text{percentage of solids} = \frac{\text{weight residue}}{\text{weight sample}} \times 100$$

(ix) Coverage by aqueous developers Aqueous developers must cover parts completely and evenly leaving neither bare spots nor lumps or heavy deposits. Areas not covered will naturally not develop indications, while a coating which is too thick can hide small indications.

The test procedure is as follows.

1. Coat half a smooth stainless steel panel (10 cm × 10 cm) with freshly prepared developer suspension.
2. Coat the other half of the same surface with used material.
3. Inspect the developer layers for any signs of failure to wet, i.e. globule formation or liquid pulling back from the edges.
4. Dry in the normal way.
5. Wash off the layers with water. The specimens should be removed equally easily.

8.2.2 Tests on penetrant materials carried out in the laboratory

The tests described in section 8.2.1 can be carried out at or even on the penetrant line. A number of other tests should be carried out from time to time – every 6 months on a busy line and at least once a year to ensure that the process is working as intended (Table 8.2). The nature of these tests is that they require equipment which should be maintained and itself audited for accuracy in a laboratory. Some companies using penetrants have suitable laboratories where such tests can be carried out; others do not, and need to arrange for samples of materials to be returned to the manufacturer for these particular tests. The tests best performed are considered below.

Table 8.2 Frequency and limits for laboratory tests for penetrant materials

Test	*Frequency*	*Limits*
Penetrant removability	Quarterly	Equal to standard
Remaining water tolerance	Weekly	50% of original
Water content of post-removable penetrants	Quarterly	Less than 0.3 vol. %
Concentration of hydrophilic remover	Quarterly	±2% of nominal concentration
Viscosity	Yearly	Within 10% of original
Specific gravity	Yearly	Within 10% of original
Fluorescence brilliance	Quarterly	At least 90% of standard

(a) Penetrant removability

For penetrant testing to be of value it must be possible to remove surface excess penetrant efficiently while leaving sufficient penetrant in surface defects to be developed to give detectable indications. Reduced sensitivity and excessive confusing fluorescent background are often found to be due to defective removal. Whether poor removal is due to changes in the penetrant or the actual remover will be indicated by the functional tests.

The removability of penetrants, whether water washable or not, can deteriorate after a tank of penetrant has stood for some time. Penetrants are mixtures and the volatility of the individual constituents varies. The more modern formulations account for this and a uniformly low level of volatility of raw materials is chosen. Disproportionate loss of the more volatile constituents will cause a rise in viscosity and will leave a mixture which is difficult to remove from the test surfaces. Insidious or sudden addition of heavy oil contaminants can cause similar effects, and persistent heating of a penetrant bath by immersion of hot parts without taking steps to keep the bath cool will accelerate the loss of volatile material. The problem is more readily noticeable when water-washable penetrants are involved, but can occur whatever removal method is used.

The test procedure again uses comparison with standard material on standard panels of grit-blased stainless steel. Panels of 16 gauge 316 stainless steel 12 cm square are prepared by grit blasting one face with grit of average size 150 μm using a gun 45 cm from the plates at a pressure of 414 kN/m^2. These panels should be kept protected.

(i) Water-washable penetrant systems

1. Cover the grit-blasted face of one panel with used penetrant and that of a second panel with standard material, and leave for the normal contact time for the penetrant.
2. Wash the panels in the normal way for the installation, i.e. spraying, immersion etc., taking care to ensure that the water temperature is normal.
3. Examine the two panels under UVA (black light) to determine whether there are significant differences.

(ii) Post-removable penetrant systems (hydrophilic)

1. Cover the grit-blasted face of one panel with standard penetrant and that of the second panel with used material. Leave for the normal contact time.
2. Rinse the panels with water in ambient UVA (black light) in the usual way.
3. Immerse in a fresh 10% solution of hydrophilic remover for 40 s without agitation.
4. Remove from the hydrophilic remover solution and rinse. Examine the panels for differences under UVA (black light).

(iii) Post-emulsifiable penetrant systems (lipophilic)

1. Cover the grit-blasted surface of one panel with used penetrant and

that of a second panel with standard material. Leave for the normal contact time.
2. Immerse both panels in standard emulsifier and allow them to drain in the normal way. Leave for the contact time used in the installation.
3. Wash with water in ambient UVA (black light), noting any difference between the removability of material from the two panels.

(iv) Solvent removal systems Solvents are very efficient removers of penetrant and comparison between standard and used penetrant by the solvent wipe method will yield very little information until the penetrant has become very profoundly changed. Where such systems are in use the difference will be more likey to be seen on artificially cracked test blocks where one half of the block is covered with used penetrant and the other with standard material. After contact time a solvent wipe and developer are employed, and differences in the number and brightness of indications will detect a deteriorated sample of penetrant.

(b) Remaining water tolerance

The nature of water-washable penetrants and emulsifiers or removers demands a degree of water tolerance or miscibility if penetrant inspection is to be available using these materials. Removers are miscible in all proportions with water and the measurement of remaining water tolerance is not appropriate for them. However, it is a very valuable measure of the state of water-washable penetrants and of emulsifiers.

The addition of water to water-washable penetrant is of course a dilution and will cause a noticeable drop in fluorescence and probably in function as indicated by the tests described above. When the water tolerance of either water-washable penetrants or emulsifiers is reached or exceeded the material will become stringy or cloudy or may form gelatinous clots and separate. Materials in such a condition should be discarded.

The test procedure is as follows.

1. Place 50 ml used penetrant or emulsifier in a graduated measuring cylinder.
2. Add small amounts of water, not more than 2 ml at a time, using a burette.
3. Stopper the measuring cylinder and invert it to achieve mixing without shaking.
4. Continue this procedure until the mixture becomes cloudy and it is not possible to read the graduations on the cylinder through the liquid.
5. Since water tolerance is very temperature sensitive steps 1–4 should be repeated with standard penetrant.

The following procedure is more accurate.

1. Weigh an empty 250 ml beaker accurately to 0.1 g.

2. Add penetrant to the beaker so that it is about one-third full and reweigh.
3. Place the beaker on a magnetic stirrer and add a stirrer bar. Adjust the stirrer so that the vortex does not meet the bar.
4. Place a light behind the beaker so that it shines through the liquid in such a way that it is not on the line of visual inspection.
5. Add water slowly from a burette with continuous stirring in such a way that each addition is dispersed before any further addition until the base point is approached. Then continue adding water dropwise until haziness persists.
6. Calculate the water tolerance as follows:

$$\text{water tolerance} = \frac{\text{volume of water used (ml)}}{\text{weight of sample (g)} + \text{volume of water used (ml)}}$$

In this calculation it is assumed that 1 ml of water weighs 1 g.

The temperature of the liquids in this test should be stabilized at $26 \pm 1\,^{\circ}\mathrm{C}$ as water tolerance is temperature sensitive.

(c) Water content of post-removable penetrants and detergent (hydrophilic) removers

The introduction of water into baths of non-water-washable penetrants is to be avoided wherever possible as the water and penetrant will separate and can cause corrosion of susceptible parts, notably steel, aluminium and magnesium.

The test procedure is as follows.

1. Place 150 ml penetrant and 50 ml xylene in a 400 ml round-bottomed flask.
2. Add some anti-bump granules and fit a condenser and a Dean and Stark apparatus.
3. Heat using a controllable electric mantle for 60 min.
4. Read off the amount of water collected in the graduated arm of the apparatus.

(d) Sundry tests

(i) Specific gravity This simple test is frequently carried out using a hydrometer, pyknometer or other means to check conformity with the original manufactured material. Differences caused by the loss of volatile constituents will only be noted when very significant changes have occurred as changes will tend to be small and specific gravity is noticeably temperature dependent.

(ii) Viscosity Viscosity measurement requires accurate temperature control as the results vary significantly with changes in temperature. Changes in viscosity can indicate quite small alterations in the constitution of the original material.

(e) Fluorescent brilliance

Fuorescent brilliance can be measured more accurately than with the on-site test described in section 8.2.1 (a) and on a busy line an accurate test should be made every 3 months. Fluorescent brilliance does not normally change very much unless there has been accidental contamination of the penetrant by cleaning chemicals such as chromates, nitric acid, alkaline permanganate or acid ferric chloride. Despite the robustness of penetrants, this test must be carried out frequently to protect the process.

The following apparatus and reagents are required: standard penetrant; methylene chloride AR; pipette filler; 5 ml pipettes; 100 ml volumetric flasks; low fluorescence filter papers (e.g. Whatman No. 4); spectrofluorimeter adapted to measure by reflectance with the primary filter passing peak energy at 365 nm wavelength and the secondary filter having a response similar to that of the human eye; drying oven set at 80 °C. The test procedure is as follows.

1. Pipette 0.2 ml standard penetrant into a 100 ml volumetric flask. Make up to 100 ml with methylene chloride.
2. Stopper the flask and mix the contents by inverting several times. Do not shake the flask or agitate it violently. Label the flask.
3. Repeat steps 1 and 2 using the penetrant under test.
4. Cut filter paper into strips of a convenient size to place in the spectrofluorimeter.
5. Take four strips for each sample and standard solution and label with a pencil. Take four blanks.
6. Immerse filter papers in the penetrant solution and remove, allowing them to drain.
7. Put all the filter papers – standards, tests and blanks – in the oven at 80 °C for 2 min.
8. Remove the filter papers and allow them to cool.
9. Measure the fluorescent brilliance of all the papers.
10. Calculate the fluorescent brilliance of the test material using the equation

$$F = \frac{T - B}{S - B} \times 100$$

where F is the fluorescent brilliance of standard, T is the reading for the test material, S is the reading for the standard material and B is the reading for the blank.

8.3 QUALITY CONTROL OF PENETRANT EQUIPMENT DURING MANUFACTURE

Penetrant equipment may be complicated or simple. As obtained from specialist manufacturers it may be standard or special. In any case it will have to be designed, manufacturing drawings will have to be prepared and then the equipment will have to be built. All phases of this process will be covered by the manufacturer's own quality control programme. This in turn will relate to various national standards. British Standard BS 5750 is an example of such a standard for the United Kingdom. The whole procedure will be subject to audits and controls to ensure that the quality standards are operating.

The quality control of penetrant lines must begin at the design stage. It is not unknown for a user to consider a penetrant line as simply a series of tanks. This is to some extent true, but it is essential for the series of tanks to be well designed both individually and as a complete operation.

Quality has been defined as the totality of features which make something fit for its intended purpose. A good design for a penetrant line allows a suitable installation to be built and avoids driers which leave parts wet and powder storm developer stations which either do not circulate the dry powder developer or allow large clouds of dust to escape every time the unit is opened after use.

Good design and good manufacture must be followed up by good installation. No task can be expected to be performed well if it must be carried out in difficult circumstances. Operators and inspectors will not maintain a high level of competence if they are uncomfortable and have to work on equipment requiring awkward manipulations or if the equipment is sited in some dark corner where the atmosphere becomes unpleasant during a working shift.

8.4 PERIODIC CONTROL OF EQUIPMENT AND ACCESSORIES

Once a penetrant line has been designed well, built to specification and installed properly in a suitable location it requires only simple maintenance (Table 8.3).

8.4.1 Efficiency of UVA (black light) lamps

UVA (black light) lamps for inspection should give an energy level of 1500 $\mu W/cm^2$ at a surface perpendicular to the axis of the beam at a distance of 30 cm from the lamp.

The following apparatus is required for testing: clamp and stand; 30 cm ruler; suitable radiometer. The test procedure is as follows.

1. Clean the filter of the UVA lamp while it is switched off and cold. If the filter is separate from the bulb, remove the filter, clean both surfaces and then refit.
2. Clamp the lamp so that it is pointing vertically downwards.

Table 8.3 The recommended frequency of maintenance checks on penetrant process equipment

Test	*Frequency*
UVA intensity	Monthly
White light from UVA lamps	Quarterly
Ambient white light	Daily
Electrical and mechanical state of UVA lamps	Monthly
Cleaning of tanks	Every 6 months
Checking tank levels	Daily
Checking driers	Monthly
Checking powder storms	Monthly
Checking general cleanliness	Daily
Checking air-line cleanliness	Daily

Note that specifications may require different frequencies of all control checks.

3. Adjust the lamp so that the radiometer detector can be placed 30 cm vertically below it.
4. Switch on the lamp and leave to stabilize for 20 min.
5. Place the radiometer detector 30 cm vertically below the lamp.
6. Place a piece of paper marked up as shown in Fig 8.7 under the detector and plot the area under the lamp where the UVA intensity is above the required minimum.
7. Take two photocopies of this plot. Retain one for the records and leave one at the line for the operator's use.

8.4.2 Checking white light emission from UVA lamps

Strictly, true UVA lamps should be invisible. However, there are very good reasons why some visible light (or white light) should be emitted. A totally invisible UVA lamp would be a serious hazard as such lamps become hot (the filters reach temperatures well in excess of 100 °C) and it is necessary for the lamps to be seen for safety. A low level of white light is also helpful in showing the outline of components, thus allowing inspectors to identify where indications are. Provided that the level of white light is sufficiently low and the wavelength is not between 450 and 600 nm the white light has little potential effect on inspection. The problem arises when the level of white light between 450 and 600 nm wavelength is enough to reduce contrast. A further problem arises in the choice of instrument to measure the level of white light. An instrument with a flat sensor which measures the energy emitted within a clearly defined waveband (typically 380–750 nm) should be used. The level of white light acceptable from a UVA lamp varies according to the level of UVA emitted. When an instrument which is sensitive to white light measures energy in the wavelength range between 380 and 750 nm the level of white light

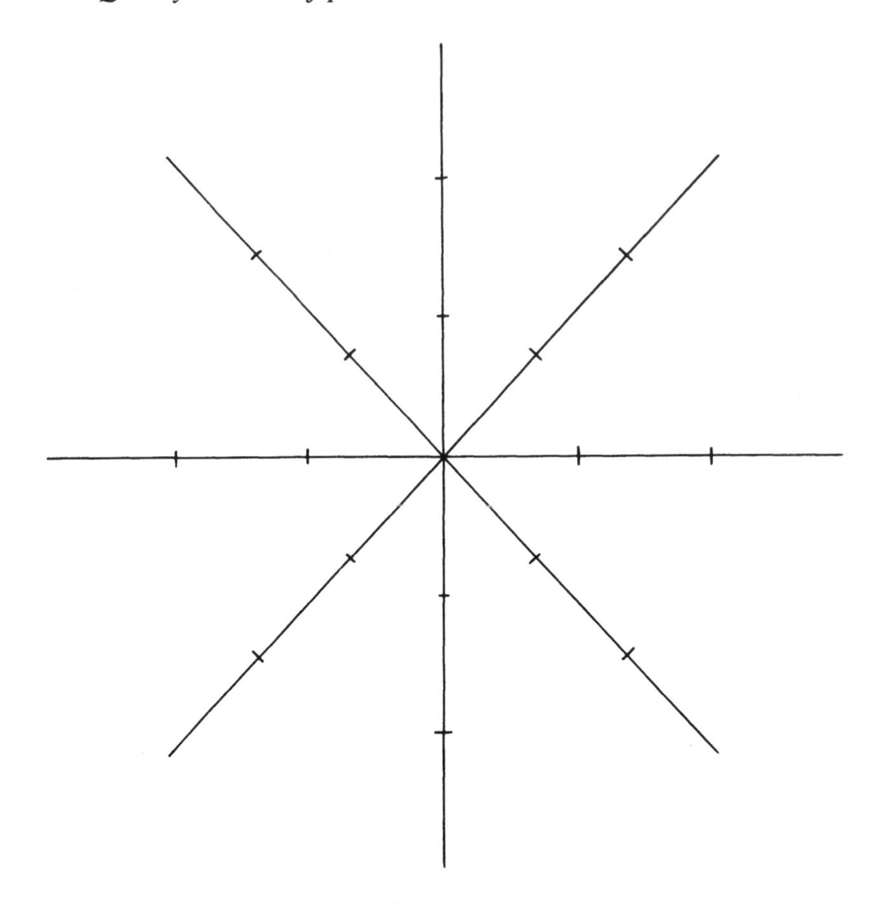

Fig 8.7 Grid pattern designed to plot UVA illumination in inspection areas

acceptable is approximately 20 lux (2 ft candles) for every 1000 μW/cm^2. When instruments measure the energy from wider or narrower wavelength bands the acceptable level of white light per 1000 μW/cm^2 will be greater or less accordingly.

A test procedure similar to that described in section 8.4.1 can be used with a white light meter replacing the radiometer and with the test taking place in conditions or ambient white light of less than 10 lux (approximately 1 ft candle).

8.4.3 Checking ambient white light in inspection areas

The level of white light in inspection areas is important whether inspection is of colour contrast indications under white light or of fluorescent indications under UVA in a darkened area. A suitable white light meter with a flat sensor measuring a wavelength band between 380 and 750 nm is ideal for this purpose. Inspection of colour contrast indications needs a minimum ambient white light level of 500 lux (46 ft candles). When fluorescent indications are inspected the ambient white light level in the inspection area (not that emitted directly by the UVA lamp) should not exceed 10 lux (1 ft candle). Note that specifications may require the use of named instruments and may require different values for these measurements. It is also important to remember that washing powders include optical brighteners to give a good white finish to laundry. It is therefore essential to avoid the use of white overalls, laboratory coats or shirts in a dark inspection area as these will fluoresce and increase the level of white light to unacceptable levels and interfere with inspection.

8.4.4 Electrical and mechanical state of UVA lamps

Lamps must be inspected for cracked filters, and any cracked filter must be thrown away and replaced. All electrical connections should be checked regularly.

Frequent changes in line voltages reduce the life of UVA bulbs for inspection lamps. Line voltage should be kept to within ±10%. Frequent switching on and off also has the same effect. Once a lamp has been switched on it should be left on for the entire shift.

8.4.5 Cleaning tanks

The procedure for cleaning tanks is as follows.

1. All tanks must be kept clean and regularly emptied, cleaned and refilled, thus preventing a build-up of contamination.
2. Unless the entire contents of a tank are to be rejected, e.g. change of hydrophilic remover solution, a period of 12 hours (overnight is ideal) should be allowed for settling.
3. The clean part of the chemical can be transformed to clean drums for re-use. This can be done with a pump, but it is easier when two drain points are provided and the clean material is removed through the upper one. The residue should be removed and destroyed.
4. Tanks should be cleaned out thoroughly using a process which will leave them clean and dry.
5. The clean material should be returned to the tanks and levels made up with fresh material.

8.4.6 Checking the levels in tanks

The level in processing tanks must be maintained high enough to allow proper processing.

8.4.7 Driers

The effectiveness of driers should be checked using temperature probes or thermometers at various points. Air mobility must also be checked.

8.4.8 Powder storm controls

Powder storms should be cleaned out regularly and the developer powder replaced. After a period of use the nature of the developer powder changes and the sensitivity of the penetrant system falls.

8.4.9 General cleanliness

All areas must be kept clean and free from penetrant chemical spills. The inspection area in particular must be kept free from developer which has fallen off parts and smudges of colour or fluorescence.

8.4.10 Air-line cleanliness

The traps on air lines must be checked and if necessary cleaned out daily, and air lines should be blown through before being used on a part. Few things are likely to enrage an operator more than a spurt of oil or dirty water across the surface of a newly processed part. If the operator is also the inspector the inspection may well be affected by the inspector's mood after such an incident.

QUESTIONS

1. Test panels and test pieces must be cleaned thoroughly immediately after use. This is because
   (a) the post-cleaning process must be checked at the same time as the function test
   (b) penetrant residues will dry out in the defects very quickly and clog them up; they are often very difficult to remove and the test piece may become useless
   (c) penetrant residues are corrosive and may damage the defects
   (d) any residues of penetrant may lead to staining of the plastic or paper in which the test panels might be stored

2. Material conformity checks carried out by manufacturers to ensure that each batch of a named product is essentially the same are
   (a) a simple functional test on the process
   (b) a few physical measurements
   (c) tests for sulphur and halogen residues
   (d) a comprehensive programme of physical and chemical tests

3. Periodic tests are carried out on penetrant materials which are in use. These are

   (a) simply daily functional tests where a test panel is passed through the installation with the first components
   (b) a daily check on timers, water pressure, temperatures etc.
   (c) a programme of tests devised to ensure that the quality of the materials does not vary during use
   (d) a check on operators' and inspectors' techniques to ensure that it does not vary

4. When assessing the daily functional test

   (a) the appearance of the test piece must match the recorded results of standard processing exactly
   (b) the appearance of the test piece must match the recorded results of standard processing essentially with allowance being made for the characteristics of lacquer developer or photographic technique
   (c) the test piece must show the outline of defects which which be well known to all the inspectors
   (d) the test piece should be processed with standard materials after the daily check, and records of the results and comparison should be made

5. If the daily functional test shows a marked loss of performance the first action taken is

   (a) to compare the performance of the used materials as a set with the same set but with one material replaced with standard material in a stepwise manner until the defective material is found
   (b) to replace all the materials which have been used immediately
   (c) to check the preparation of the components for possible interference with the process
   (d) to check the penetrant for contamination and loss of brilliance

6. When solutions of detergent (hydrophilic) remover are used the concentration must be checked. This can be carried out at the installation by use of

   (a) a thermometer
   (b) a refractometer
   (c) a psychrometer
   (d) a hydrometer

Note: There is only *one* correct answer to questions 1–6. Tick your choice and check it with the correct answers on p. 215

9

Specification and documentation

9.1 INTRODUCTION

Penetrant testing, in common with all forms of non-destructive testing and evaluation, is a part of the quality control procedure. As such it must be carried out in a controlled way so that the results can be useful. In order to achieve this the work must be repeatable and traceable. One of the most satisfying aspects of penetrant testing is that the results can be seen. However, it is very difficult to record these results. All methods of recording the images of tested components are very slow, as in the case of photography or transfer lacquer, or very expensive and restricted in application. The various forms of scanning devices which are available are all well adapted to the inspection of large numbers of a single component or are capable of inspecting a range of components of simular form. While the possibility of designing and building a device for inspection of a very wide range of geometries is possible, the cost and difficulties of application are problems which await a solution. In view of this aspect of the penetrant method it is clear that all efforts must be made to ensure that the processes are employed in a controlled way.

9.1.1 Specifications

There are a large number of specifications which are designed to control the use of penetrant testing. These specifications fall into two groups: one type of specification deals with the manufacture of equipment and materials to be used in the process and the other deals with the actual application itself. Some specifications contain elements or aspects of both controls. It is very important to have a clear understanding of the two aspects of control by specification.

Initially it is important to ensure that the equipment to be used, if any, and the penetrant materials have been manufactured to a recognized standard. It is then equally important to ensure that the penetrant processing is carried out to a recognized standard, that the equipment is maintained to the original standard and that the penetrant materials are still performing to the original standard. The first stage can be dealt with at the stage of commissioning the facility. The equipment can be certificated, and certificates of conformity for the materials can be obtained from the manufacturer. Once the penetrant facility is in use it is essential to maintain a record of the control checks on it. Where the control checks are contained in a separate specification this must be applied rather than the original document. Where the installation of the process and its operation are both contained in one document it is important to ensure that the records of controls on operation refer to the correct clauses and paragraphs so that confusion can be avoided. Many of the control checks on the penetrant processes can, and indeed should, be carried out at the installation. However, some control checks involve sampling materials and laboratory tests which are carried out away from the site of use.

In this second case it is very important that the correct specifications, paragraphs and clauses are called up when tests are required if the documentation is to be effective and pass an auditor's scrutiny.

In addition to the specifications that have been prepared by national bodies there are specifications which are prepared by companies for control of their own production and the production carried out by their suppliers. The aerospace industry offers the largest number of these at present. Other industries such as power generation, also issue similar documents. Naturally these specifications are prepared to cover the individual needs of the particular authority. In some cases separate documents exist to cover new manufacture and overhaul in recognition of the different practical problems of these two aspects of the application of penetrant testing and the different types of defects which are likely to be found.

Other specifications have very special requirements which are not concerned with penetrant processing as such but make requirements for levels of certain chemical elements. These requirements are designed to ensure that traces of elements which may be metallurgically objectionable are kept to a minimum. Some process specifications also include requirements for maximum levels of elements such as sulphur and the halogens.

Despite the large number of specifications designed to control the application of penetrant testing is still quite common for this method of NDT to be applied with none of the standard specifications. In such cases it is essential that some form of internal specifications can be made to control the work if maximum benefit is to be obtained. Reference to the national specifications is often a very useful starting point. However, these are necessarily written in a very broad way. An appropriate process or a choice of processes can be taken from such a specification and written into the internal document, and the details of the process steps, such as penetrant contact times, water washing times etc., can be set out. Some documents of this type refer to other specifications, such as British Standard BS 6443 or European specification EN 571, by name. While this route has some attractions it also has some disadvantages. First, many penetrant specifications prepared by national or professional bodies are very broad in concept and are open to very widely varying interpretation so that individual control can be lost, and, second, such specifications are revised periodically so that any documents derived from them must be kept current. It is therefore more useful to prepare a procedure to control the individual application of the penetrant method. Such a document should cover initial procurement, processing, periodic controls, and approved materials and accessories.

Ideally, penetrant materials which are acceptable for use in the various penetrant processes are identified in a qualified products list which can be revised separately from the remaining documents. This avoids the complication of revising an entire document when it becomes necessary to add new products or to delete products which were once approved and either cease to be available or are no longer considered suitable for inspection. When penetrant materials are approved by name, the complete sequence of penetrant, remover and developer should be specified in the list of qualified products.

When accessories or instruments, such as light meters, UVA meters or refractometers, are referred to it is more useful to describe the function of such accessories or instruments in the main specification and list specific models and suppliers in an appendix which can be maintained separately. This allows control of problems such as discontinuation of the availability of a piece of equipment and changes of addresses of suppliers.

When specific physical or chemical measurements are required, as is the case with many control methods for materials, these should be stated in full. The use of standards from other professional bodies is attractive until such methods are changed. This problem is not resolved either by specifying the current method according to a third party specification or by stating a date or edition of that specification. The first route leaves the penetrant specification open to uncontrolled changes by a third party who may not be aware of the specific use of their method. Their revision may leave a method which is acceptable for controlling the quality of the penetrant process materials or it may not. The objection to the second approach to using third party specification is that they cease to be available as soon as they are revised. In summary a specification must as far as possible fulfil the following conditions:

1. properly defined;
2. easily understood;
3. independent of other documents;
4. organized such that one section can be revised without involving the whole document;
4 openly available to all justifiably interested parties.

9.1.2 Associated documents

Specifications can give details of the manufacture of penetrant materials, the design and manufacture of penetrant processing equipment, the application of the materials at all stages of the processes, a list of approved materials complete with classification and a description of methods to maintain the equipment and material to an acceptable standard. It is inappropriate for penetrant processing specifications to contain details of sentencing components. Sentencing, whether of finished components, part finished components or raw material stock, must be covered by a quality assurance document. Such a document must set out acceptable limits which may range from no indications to limits on dimensions of indications, e.g. for cracks, forging defects etc., to the density of indications, i.e. the number of indications allowed in any given area as might be the case for porosity. Preparation of quality assurance documents should be the result of discussion between the design engineers and the NDT authority. The design engineers know what level of performance the components must be capable of; however, their appreciation of the capabilities of penetrant testing or any other NDT method may not be so complete. Sensible discussion between these two authorities should lead to a solution which can be applied in a practical way.

There is no hard and fast rule which can be laid down as to how penetrant processing can lead to the release of acceptable materials and components for further processing or use. However, the elements must include the following:

1. controlled processing;
2. the method of processing allowed for individual components or material;
3. a quality acceptance standard;
4. clear documentation.

9.2 DETAILS OF PENETRANT PROCESSING SPECIFICATIONS

Whether a specification is designed to cover the most critical component of an aero-engine or a weld in a storage tank for a non-hazardous chemical, the basic elements set out in this section must be considered and dealt with. It is also important to note that the actual detailed parameters may vary very significantly depending on whether the process is manually or automatically operated.

9.2.1 Control of personnel

At one time the training of operators and inspectors for penetrant testing was very variable. In some industries and some countries a suitably serious view has been taken of this and systems for training personnel have been available for many years. In recent years application of these standards has become much more widespread. This is a development which is to be welcomed, and the day cannot be far off when unsupervised learning of the technique 'on the job' is a thing of the past. Apart from training, other aspects of personnel which must be covered in a specification include visual acuity and colour vision.

Uncorrected eye defects present self-evident problems for the value of any visual inspection process. While colour blindness may not necessarily be a bar to operators or even inspectors, as in colour contrast penetrant testing the red of the indication will appear black on white rather than red on white and in the case of fluorescent penetrant testing indications will appear bright white on a dark ground. However, a colour blind inspector may find difficulty in differentiating between traces of dark material surface showing through a developer layer when using the colour contrast process or in identifying pieces of cotton and other material which may fluoresce under UVA when a fluorescent process is used. At least, any defective colour vision should be known and accounted for.

Two other aspects of the control of inspectors are dark adaptation and the length of time spent in inspection without interruption. When people move from an area of normal or high ambient white light to one of low ambient white light, as happens when an inspector enters darkened inspection booth, two important changes occur. The first is opening or dilation of the pupils of the eyes so that maximum light is collected, accompanied by an increase in light/dark vision as opposed to predominantly colour vision, and the second is the shift of peak visual sensitivity for colour from the yellow of daylight vision to a lower wavelength corresponding to yellow green. Both these changes take time, and operators need to wait after entering a dark booth before starting to inspect components or materials which have been processed using one of the fluorescent penetrant processes. This provision is necessary only for fluorescent

Table 9.1 Personnel requirements which need to be stated in a penetrant processing specification

Aspect	*Comments*
Training and experience	Formal training to a recognized requirement followed by a period working under the supervision of an experienced and qualified colleague
Visual acuity	Checked against professional standards at least every 12 months and any defects corrected
Colour vision	Checked at the same time as visual acuity; any defects noted
Dark adaptation	Fluorescent penetrant processes only; allow at least 2 min after entering a dark inspection booth before starting any inspection
Inspector fatigue	Do not continue uninterrupted inspection for more than 2 hours, a 15 min break will restore the operator to acceptable performance

The figures in this table are recommended minimum requirements. Experience of specific applications is likely to indicate different figures.

penetrant inspection methods. The second requirement limits the length of time that an inspector should work uninterrupted and this applies to any inspection process. While the human mind is very adaptable it is clear from many organized studies that attention starts to wander and concentration lapses after a period of time. The period allowed before a rest varies according to the severity of conditions and the level of concentration needed for a specific inspection. It is generally acknowledged that the powers of concentration of all people for a repetitive task begin to deteriorate after 2 hours and this limit is often used. The break before restarting may be quite short, as little as 15 min. The personnel requirements which should be covered by a penetrant processing specification are listed in Table 9.1.

9.2.2 Manufacturing stage

Although the manufacturing stage is dealt with more precisely in the individual documentation of a component and the preparation of surfaces requires specifications for the various available methods, these points should be covered in any specification designed to control penetrant processing.

Table 9.2 Requirements dealing with component preparation

Aspect	Comment
Manufacturing stage	After processes which may induce surface cracks and before processes which may mask cracks or interfere with penetrant processing
Cleaning surfaces	Remove contaminants according to cleaning specifications; ensure that surfaces are clean and dry before applying penetrant; ensure that the temperature of the surfaces is appropriate

In general terms penetrant testing should be carried out after any process which could cause surface-breaking defects and before any process which may close up such defects or interfere with the penetrant process (Table 9.2). Manufacturing processes such as casting, forging, machining, grinding, spinning, heat treatment, stamping, welding etc. can cause defects in surfaces. Shot peening, sand blasting and polishing can close surface defects completely or restrict access to them, and some treatments such as anodizing, chromating and other surface conversions often cause problems in the penetrant process.

9.2.3 Preparation of surfaces

Surfaces must be clean and dry before application of penetrant (Table 9.2).

In the most commonly used penetrant processes the surface temperature is between 5 and 40 °C (40 and 105 °F). In some cases it may be necessary to test surfaces at temperatures which may be significantly above or below this range. In such cases specially formulated penetrant materials must be specified.

The actual method of cleaning surfaces in preparation for penetrant testing depends on the contaminants on the surface. Where use of a volatile organic solvent is adequate, the possibility of contamination of the penetrant on application is small. However, when paint strippers, carbon removers, etchants and other more aggressive chemical mixtures are used, it is essential that all traces of these materials and any water used in final rinses is removed before penetrant is applied.

9.2.4 Penetrant application and contact time

It really does not matter how penetrant is applied to the surface to be tested as long as the entire area is covered and remains covered for the specified contact time. It is important to know how the penetrant is applied as different methods of application leave different thicknesses of penetrant on the surface and this influences the severity of penetrant removal. A number of methods are used (Table 9.3).

Table 9.3 Penetrant application requirements

Aspect	Comment
Application	
Brush or swab	Brushing mandatory for application of thixotropic penetrant; ensure that the gel is broken down on application
Immersion	Ensure that drainings from the component do not return to the tank unless they do so directly
Spray	Observe sensible health and safety precautions
Electrostatic spray	Observe health and safety precautions; ensure that component surfaces are completely covered
Flow on pour on	Ensure that drainings from the component do not return to any reservoir unless they do so directly
Fog	Ensure that penetrant is restricted to its station
Contact time	Use the contact time given for the individual component; this will be a minimum of 10 min

(a) Application of the penetrant

(i) Brushing or use of a swab Brushing and swabbing are in widespread use for local application. It is very useful when certain special processes are used. These include application of thixotropic materials and high temperature materials which can be used at temperatures up to 200 °C. When thixotropic penetrant is used a brush is essential as it is necessary for the gel to be broken down by the stippling action if the potential sensitivity of the process is to be achieved. Brushing or swabbing on are the preferred methods of application for the materials used in the high temperature colour contrast penetrant inspection process. Materials that can be applied to surfaces which are as hot as 200 °C (392 °F) are available for colour contrast penetrant processing (similar materials are not available for the fluorescent penetrant process) and, although such materials can be applied by spray or immersion, use of a brush or swab is generally more practical.

(ii) Immersion Immersion remains a common method for application of penetrant. Specifications should indicate that immersion tanks are large enough to allow complete components to be immersed if this method of application is chosen. It should be noted that drainings of penetrant material should only be allowed to return to the penetrant tank directly from the component while it is suspended above the reservoir. Penetrant drainings which are collected on a drain tray should be disposed of as waste since, over a period, the quality of such material cannot be trusted. Immersion tanks for penetrant should also be equipped with dual drainage outlets to allow periodic cleaning. A drain outlet 10 cm (4 in) from the base in addition to one in the base allows the clean penetrant to be drawn off easily and the heavily contaminated material at the bottom to be separated easily during this control check. It is good practice for such tanks to be equipped with lids and, if several penetrants are incorporated in a composite line, the most sensitive penetrant should be placed in the first tank so that in the case of accidental contamination through drainage the defect-finding capability of the lower sensitivity penetrants is increased rather than decreased. Under no circumstances should tanks of colour contrast penetrant and fluorescent penetrant be combined in one installation.

(iii) Conventional spray application (including aerosols) Conventional spray application includes the use of spray guns and aerosols, of which application by aerosol spray is the more common. Very few specific points need to be raised except those affecting safety. Sprays of all materials can be a nuisance and failure to take account of wind direction when using aerosols of colour contrast penetrants in the field can literally lead to red faces. Care must also be taken not to spray such materials near sources of ignition. Despite the fact that modern materials have flash points above 100 °C (212 °F), the finely divided material of a spray mist will catch fire quite readily.

(iv) Electrostatic spray application Electrostatic spray application has justifiably become popular as a flexible and economical method of applying penetrant. The major specification requirement is to check that the component is actually covered since when properly applied the penetrant layer is so thin that, in the case of a fluorescent penetrant, the layer is invisible. In such a case this can be checked readily enough using a UVA lamp. When colour contrast penetrant is used, the colour is strong enough for this check to be carried out in good white light.

(v) Flowing on or pouring on Flowing on or pouring on may involve total loss of penetrant material or re-use by pumping material to a hose or an array of nozzles whereby the penetrant drainings are collected and recirculated through a pump. Where recirculation is involved, drainage must be directly into the reservoir, and fairly frequent control checks on the penetrant are advisable to guard against cumulative contamination.

(vi) Fog application Fog application has recently gained popularity in the United States. It is difficult to consider how such a method could be used in a manual process; however, it may be attractive in automated installations. The most important specification requirement must be to ensure that the penetrant stays in its own station.

(b) Penetrant contact time

Penetrant contact times in specifications are usually stated as minimum and maximum values. More precise times are given in procedure documents for individual applications. It has been indicated elsewhere that penetrants need time to work, and on some materials experiment will show the need for contact times of 30 min or more. Despite commercial claims to the contrary, contact times of less than 10 min are not useful. Certainly, some defects can be found in some materials when contact times of as little as 2 min are used and the indications are often enough to lead to rejection of the component. If a longer penetrant contact time is used on components which do not have coarse easily indicated defects, the possibility of finding indications of much finer yet equally important defects becomes a probability and the cost of 8 min must be weighed against the direct and indirect costs of a subsequent failure. In fact the saving of 8 min is illusory if more than one batch of components is to be tested as, after the initial batch of the day has been tested, operators will use the penetrant contact time in other parts of the process. Actual minimum penetrant contact times should be set for each problem. The indication of maximum penetrant contact times is largely historical, as older penetrants were volatile and dried out during this period, leading to difficulties with removal. Nowadays it can still be argued that a maximum penetrant contact time is needed to ensure some comformity in processing. In practical terms the components can be suspended over a reservoir of penetrant during contact time, in which case the drained material can be collected for re-use. However, in many cases the throughput capacity of the installation is increased by having a separate area where components awaiting the next stage are parked. In this case penetrant drainings should be collected and sent to waste as the quality of such material cannot be controlled.

9.2.5 Penetrant removal

The removal of surface excess penetrant at the end of penetrant contact time is achieved in a number of ways. When a specification needs to cover several of these it must, of course, detail all those required. However, when only one method is required, only that method needs to be indicated. When several methods must be detailed this can be done either by listing the details of penetrant removal with each type of process and assigning a code identity to each or by presenting it as a composite section in the specification. Either approach is effective provided that the detailed information is given. In this section each of the methods is discussed (Table 9.4).

Table 9.4 Penetrant removal requirements

Aspect	*Comments*
Water wash	Standardized water wash requirements for all stages (where this step is needed) avoid confusion Specification must indicate • Method of application • Spray pressures (where spray is indicated) • Temperature • Minimum distance between spray heads and surfaces • Whether air assistance is allowed or forbidden • Maximum time allowed for wash • Avoid mixing of dissimilar metals
Solvent removal	• Liquid or vapour • Time of exposure to vapour
Detergent remover solution (hydrophilic removers)	• Method of application • Concentration in water • Maximum contact times • Maximum temperature
Emulsifying agents (lipophilic removers)	• Method of application • Maximum contact time • Operating temperature range • Special provison to stop emulsification
Special removal methods	• Define all possible variables as closely as possible

(a) Water washing

Water washing is necessary in four instances in the commonly used penetrant processes and in several of the more unusual procedures. Where water washing occurs several times in a specification there are significant reasons for using standard conditions for the procedure. The most obvious of these is a lack of confusion. Irrespective of whether water washing conditions are standardized throughout or are varied according to which water wash is in operation the following points must be covered.

1. Application – spray, immersion or both.
2. Spray application:
   (a) pressure – a maximum must be given;
   (b) temperature – a range of temperature must be stated;
   (c) minimum distance between the spray head and the surface must be given;
   (d) is air assistance to be allowed or not?
3. Time – a maximum time for any part of the surface to be washed with water must be stated

It is also important to specify the equipment clearly. Where immersion water rinsing is specified, the water reservoir will become contaminated and some control must be exercised on the level of such contamination that can be tolerated. Specific attention must be paid to any metals which may be in contact with components. Many baskets and carriers are made of stainless steel, and hooks for suspending large components are often made from carbon steel. A combination of water and dissimilar metals is the basis of an electrochemical cell. Many aluminium, magnesium and other valuable components have been reduced to scrap as a result of unfortunate contacts in penetrant wash stations. Specifications should require that such components are kept in contact either with non-metallic carriers or with carriers made from similar metals.

(b) Application of solvent removers

The loose term 'solvent removers' is taken here to refer to volatile organic solvents. Specifications covering this method must include the method of application. This may be as a liquid or as a vapour.

- Application as a liquid – this must never be direct but by the 'wipe technique' and must be described fully. The identity of permitted solvents must be given, e.g. non-halogenated or halogenated.
- Applications as a vapour – this method of application excludes non-halogenated volatile solvents as these would present an unacceptable fire risk if they were used in this way. Since organic solvent vapour is an aggressive penetrant remover, timing of this procedure is important and must be set out in a specification.

(c) Application of detergent remover solutions (hydrophilic remover)

Hydrophilic removers are supplied as concentrates and so specifications must indicate the concentration at which they are to be used. Other factors which must be specified are as follows:

- Method of application – immersion, spray or foam
- Maximum contact times
- Maximum temperature – at temperatures above 40 °C (105 °F) the activity of these chemicals changes

(d) Application of emulsifying agents (lipophilic removers)

Lipophilic emulsifiers are supplied ready for use and specifications must indicate the following:

- Method of application – immerse and drain or spray and drain
- Maximum contact times – this is much more critical than for detergent removers
- Temperature range – the speed of action of this type of remover will roughly double for an increase of 10 °C (18 °F) in temperature

Some emulsifying agents are very rapid in their action and a water spray, whether manual or automatic, may not be capable of stopping the removal action uniformly over the complete surface of a component. In such cases special provision such as an immersion water bath may be specified.

(e) Special removal techniques

From time to time circumstances arise where special techniques for penetrant removal have been devised. Two examples are the use of corn husks and the spraying of fruit stone grit on to the surface after penetrant contact time followed by a water spray rinse. It is very important that any special removal technique be defined as closely as possible in order to maintain control over the process.

9.2.6 Drying components

When volatile organic solvents are used to remove surface excess penetrant, the components can be left to air dry. It is advisable to check that no reservoirs of solvent remain trapped in areas of components which have a complicated shape. The final stage of many penetrant removal processes is a water wash. At some stage this must be dried. The point at which drying takes place depends on the type of developer used. If a dry powder or a non-aqueous liquid developer is to be used surfaces must be dry before the developer is applied. When an aqueous developer is used this is applied to the wet surface and drying follows application of the developer.

In each case drying must be carried out using warm mobile dry air. Specifications must indicate the permitted method of drying, e.g. oven or warm air blast, the maximum temperature and the length of time allowed for drying (Table 9.5). The temperature indicated would ideally be that at the tested surface. However, this is not practical and the temperature of the applied air is more readily controlled.

Table 9.5 Drying specifications

Aspect	*Comments*
Drying method	Warm air oven with recirculation Warm air blast
Maximum temperature	Temperature of the air is more readily controlled
Maximum time allowed for drying	For most components up to 100 kg this is normally 10 min; components of greater mass will need longer drying times

9.2.7 Developer application

Specifications must indicate which of the various types of developer are allowed, the permitted methods of application and the minimum length of time that developer should remain on the surface before inspection can take place (Table 9.6). The precise details of these factors will vary according to the type of developer.

(a) Dry powder developers

Dry powder developers are of interest only when specifications are prepared for the control of fluorescent penetrant testing as this type of developer does not provide a suitable background for colour contrast processes.

Methods of application of dry powder developers should be such as to avoid the possibility of contamination and account should be taken of the fact that the fine dust from such materials can be a nuisance. Approved methods include storm cabinets, electrostatic sprays, fluidized beds and, where small numbers of components are concerned, dusting on through a wire mesh or with a rubber pear puff applicator. Dry powder developers should be left on components for a minimum of 10 min before inspection is allowed.

(b) Non-aqueous liquid developers

Non-aqueous liquid developers are suspensions and as such must be agitated thoroughly before application. The only permitted method for application is spraying. Specifications should indicate the amount of developer which should be deposited on surfaces. In the case of colour contrast penetrant processes

Table 9.6 Developer requirements

Aspect	*Comments*
Type of developer allowed	Dry power Non-aqueous suspension Aqueous solution Aqueous suspension
Method of application	Storm cabinet Electrostatic spray[a] Dust on } Dry powders only Spray on Immersion Curtain Brush on } Liquid developers only
Concentration (aqueous developers only)	Normally in accordance with manufacturers' instructions
Drying temperatures (aqueous developers only)	In accordance with manufacturers' instructions or 60–80 °C (140–178 °F)
Developer contact times	
Minimum	Needed for all types of developer
Maximum	Needed for non-aqueous suspension developers and advisable for aqueous suspension developers. Many specifications indicate maximum values for all types. If inspection is delayed beyond 30 min components must be protected from UVA during this time

[a]Electrostatic spray application can be used for non-aqueous suspension developers. However, in normal circumstances the advantages are minimal and the extra cost involved cannot be justified.

the developer should be applied to give a uniform white covering which just obscures the surface. When this type of developer is used in the fluorescent penetrant processes the developer should be applied in such a way that the surface is still discernible through the layer.

Non-aqueous liquid developers generally work more quickly than dry powder developers and inspection can take place 5 min after application. Some of these processes lead to heavy bleed-out of penetrant which can cause the indications to become diffuse in quite a short time. In view of this, a maximum allowed time before inspection should be specified as well as a minimum.

(c) Aqueous solution developers

Aqueous solution developers are normally supplied as dry powder concentrates to be made up for use in water. Specifications must indicate the concentration range allowed, the method of application, the drying temperature and the minimum length of time to elapse between application of the developer and inspection.

The concentration of such developers depends on their formulation, and specifications should indicate that the working solutions should be made up in accordance with the manufacturer's instructions. Application can be by immersion, curtain or spray – in fact, any method which ensures a complete coverage of the surfaces. Since this type of developer must be dried after application the drying temperature must be considered. The manufacturers may recommend a drying temperature range and in such cases their recommendations should be followed. In the absence of manufacturers' instructions components should be dried in hot moving air (at a maximum temperature of 80 °C (178 °F) for a maximum of 10 min.

The minimum time before inspection must also be stated. In view of the temperature used to dry the developer indications form quite quickly and inspection can usually take place after developer contact time of 7 min. Indications formed in layers of developer of the aqueous solution type are normally quite well defined and do not bleed heavily even after a long time, and so there is no practical need for a maximum developer contact time to be stated.

(d) Aqueous suspension developers

Most aqueous suspension developers are also normally supplied as powder concentrates to be mixed with water. Once prepared they must be agitated continuously during application. A few are permanent colloidal suspensions. Specificatons must give details of the concentration to be used, normally in accordance with the manufacturer's instruction, the method of application, which may be by immersion or spray, the drying temperature, either in accordance with manufacturers' instructions or any requirements of the specification, the minimum and possibly maximum developer contact times before inspection can start, and, if a maximum time is given, the time within which inspection must be completed. A note should be included to indicate the care needed in applying such developers to ensure that an even coating is achieved, particularly in keyways and at changes of section.

It should be noted that some practices allow a redevelopment technique whereby an indication may be wiped with a volatile solvent on a brush or swab. This can have no effect, cause the indication to disappear momentarily and then reappear or simply cause it to disappear. When the first two results are seen it is assumed that the indication is of a genuine discontinuity. Unfortunately, if the indication disappears it is not justification for the assumption that it was false or irrelevant. In this case the component should be cleaned and reprocessed.

9.2.8 Inspection

Most components are inspected by eye and it is likely that this will remain the case for some years to come.

(a) Human eye inspection

When inspection is carried out by eye specifications must set down certain minimum requirements and these will depend on the type of penetrant process (Table 9.7).

Table 9.7 Requirements for inspection by human eye

Aspect	*Comments*
Illumination	Minimum white light levels for inspection of surfaces processed by colour contrast penetrant methods
UVA (black light)	Minimum levels of UVA at inspected surfaces to be stated
White light (ambient)	Maximum allowed ambient white light levels to be given; guard against transient sources, e.g. fluorescent white laboratory coats
White light (from UVA lamps)	Maximum allowed level expressed as a ratio with the UVA emitted
General comfort	Low noise levels Few interruptions Good ventilation Adequate space Seating may be required

When colour contrast penetrant processing is used, a minimum level of white light must be specified. In some instances inspection will take place in the open air and on dull days in high northern or southern latitudes the natural daylight in winter may need reinforcement from an artificial lamp. Similarly, in those countries or areas where there is a pronounced period of dusk care must be taken to ensure that enough natural light is available when inspection is carried out early in the morning or during the evening. When fluorescent penetrant processing is carried out the inspection area must be darkened, and the maximum level of allowed ambient white light stated. Minimum levels of UVA at the surface to be inspected must be given. UVA lamps emit some white light and this level must also be specified. This value is best given as a ratio of white light energy to the UVA rather than as an absolute value. In all requirements for measuring white light or UVA the equipment to be used must be clearly defined and the methods for its use unambiguously stated.

All inspection areas must be as free from distraction as possible, and this applies to noise and interruptions as much as visual distractions. Transient distractions such as the wearing of white laboratory coats must be avoided as these invariably fluoresce strongly and raise the levels of ambient white light to well above the allowed maximum. As far as possible inspectors must ensure that reflections from shiny surfaces are avoided. The comfort of inspectors with regard to space, seating and ventilation should also be accounted for.

(b) Automatic inspection

For many years it has been an objective of the penetrant manufacturing industry and the users of the process to release inspectors from their repetitive task which, in the case of fluorescent penetrant inspection, takes place in a dark and frequently uncomfortable booth. Some significant success has been achieved, most notably when large numbers of essentially similar components are to be inspected which are of fairly simple shape and the indications which must be detected are also of simple shape.

Any automatic inspection system must acquire a total image of the processed surfaces, compare this image with examples of acceptable items and on the basis of this make an accept/reject decision. A third decision of pass for manual overcheck can be introduced, but when this is done the process is no longer automated inspection but automated assistance to human inspection in that it is simply reducing the number of components which the human inspectors actually have to look at. At present any specification to control automatic inspection should be written individually for the project in hand.

9.2.9 Post-cleaning

Post-cleaning of processed components before they are passed for use, storage or further manufacture is very important. It is essentially a separate process and as such is best dealt with in a specification document designed for it. In some cases the removal of penetrant residues and traces of developer is included in a penetrant specification for convenience, but this is best avoided as it reduces the accuracy of the control sought by preparing specification documents.

9.2.10 Control procedures

The original equipment for a penetrant processing installation should be purchased against a general specification or, if a special installation is required, a specification should be prepared for that particular equipment. The penetrant chemicals will also have been purchased against specifications issued to manufacturers and certification will have been obtained. Control procedures must be operated to ensure that the installation and the chemicals continue to work in the same way and to the same standard throughout the life of the installation and these control checks must be documented.

(a) Controls on equipment

Specifications should contain instructions which ensure that the equipment used in maintained in a controlled way and that such control is documented. The equipment must be checked for the following:

1. general cleanliness;
2. tank levels;
3. the cleanliness of any compressed air lines which are used;
4. operation of the processing units including handling equipment, spray apparatus, powder storm cabinets and any other mechanical parts;
5. electrical maintenance;
6. where appropriate, tanks should be cleaned out

The length of time between repeats of such control checks may vary from daily to 6 monthly and should be set out in any specification.

(b) Controls on the penetrant materials

Methods for controls of penetrant materials are given in some detail in Chapter 8. Specifications should indicate methods for testing the materials, the acceptance limits and the frequency of each test. The objective is to ensure that the penetrant process continues to give similar results to those achieved when the materials were new within an accepted variation.

(c) General performance check

The general performance check is very important and must be carried out using a clean test piece – either an artificially cracked piece or a known defective sample of the components under test – every time the penetrant installation is used at the beginning of the day or shift. Any change in the performance of the process must be investigated further. By substituting each of the materials from the installation in turn with unused material (a retained sample) the defective step or steps can be found and corrected. The fluorescent brilliance or colour intensity of the penetrant should be checked regularly.

(d) Penetrant removability

Pentrant removability must also be checked regularly. Many test pieces have areas designed to show whether the removability of the penetrant is satisfactory and some control is achieved in this way. From time to time it is advisable to check the removability of the used penetrant with that of the retained sample.

(e) Water content of penetrant

Penetrant does not normally contain water. However, water can get into penetrant which is stored in open tanks by splashes from nearby wash stations, from wet components being immersed in the penetrant or from condensation dripping into the bath from overhead structures. Despite the fact that none of these things happen on a well-designed and properly managed line specifications should require proof of this. Water-washable penetrants may fall out of specification owing to loss of water tolerance and all types of penetrant may become corrosive to components if sufficient water is added to them.

In addition to the above the following should be specified: temperature of the wash water; temperature and concentration of the hydrophilic remover solution; temperature of the emulsifying agent. The following checks must be made on the developer. These will depend on the type of developer which is used.

- Contamination – all types except non-aqueous liquid suspensions
- Concentration – all types except dry powders (this check is not necessary when non-aqueous suspensions are used in aerosols)
- Wetting ability – aqueous developers only

The frequency at which the checks listed here should be carried out is suggested in Table 8.2. However, individual specifications vary from the frequencies suggested in Table 8.2. When working to a specification the requirements of that document must be followed.

9.2.11 Control of illumination and ultraviolet lamps

(a) White light illumination (colour contrast penetrant testing only)

Once an inspection area has been set up where the level of white light meets or exceeds the level required by the specification regular recorded measurements must be made to ensure that the original standard is maintained. The method used for measuring the light level must be given and the equipment described in detail.

(b) Level of UVA (black light) (fluorescent penetrant testing only)

The level of UVA (black light) must also be controlled so that the conditions of inspection remain in specification. As for white light the method of measurement and the equipment to be used must be described fully,

(c) Levels of white light in inspection booths (fluorescent penetrant inspection only)

This is measured using the same method as that described in section 8.4.3.

(d) Levels of white light emitted by UVA (black light)

The method for measuring this must be specified very clearly and the equipment to be used clearly defined. The method must also take account of white light from sources other than the UVA lamp (black light) as this instrument will detect this in addition to any such illumination direct from the lamp.

Instruments used for measuring light or UVA must themselves be calibrated periodically.

9.3 INDIVIDUAL PROCEDURES

As there are more than 100 ways of combining penetrant materials to form a process which will give results, it is necessary to assign an individual penetrant process to each component. The factors taken into account when making the choice are discussed in detail in Chapter 6.

The identification of the processes can be included in the specification where the various penetrant processes are classified and individual products assigned to the classes and levels of potential sensitivity. In this way a component or material can be designated according to the type of penetrant process and the level of potential sensitivity by means of a simple code. Such codes can be used on drawings and all the paperwork or other records which follow the component or material through its life.

Another approach is the use of an NDT technique card. Such a document must contain the following data set out in a clear format.

- Part number and description of the component
- Origin of the component/material
- The identity of the component/material, e.g. the type of engine where a turbine blade will be used
- The issue of the technique card
- The date of issue of the document
- The penetrant specification which controls the technique card
- Other NDT techniques which are to be used
- The penetrant process to be used
- The quality assurance to be applied
- Signatures of the authorities approving the document

Whichever of these approaches is adopted, the penetrant testing can be controlled and the value of the work established by its being traceable.

9.4 QUALITY ASSURANCE STANDARDS

There is little point in carrying out any test without having a quality standard against which to measure the results. The problem with penetrant testing is that the very satisfying visual result is not easily quantified. The normal way in which a quality acceptance standard is written for penetrant processing is to require one of the following:

- No indications at all
- No indications greater than a stated length or diameter or both
- No more than a stated number of indications within a specified radius

These requirements can all be combined in a single quality assurance standard where different areas of one component are required to have different levels of integrity.

The quality assurance standard cannot be prepared solely by the NDT department because the decision as to what level of integrity each component, structure or material must have to be fit for its purpose involves a range of engineering and material science experts. It is a document which must be applied without exception when data for sentencing tested surfaces are collected.

9.5 CONCLUSION

In this chapter the aim is to indicate the origin and importance of specifications, procedures, quality acceptance standards and the need to document the whole process. Some specifications will contain details of all the probable methods of applying penetrant processes and inspecting components, whereas others may refer to a single procedure. Each approach is equally valid provided that the specification covers the inspection requirements. In the absence of a specification, procedure and quality acceptance standard, a set should be prepared. The proper order is the agreement of the acceptance standard with the authority responsible for the integrity of the material, component or structure, followed by choice of a suitable penetrant process as outlined in Chapter 6 and preparation of a specification. This work must be followed by a controlled operation of the system. It is difficult to see much value in carrying out any inspection without a documented procedure. Perhaps the manufacturer may sleep easier at night for carrying out inspection, but some day someone is going to demand proof of these controls. When writing such documents always list approved materials in a separate appendix which can then be revised without the problem of rewriting the whole document. This approach allows opportunities for wider choice, and when individual products are no longer available the specification does not become out of date. Finally, when a specification and the associated documents are provided by a customer, they should be read thoroughly and advice sought on any points which may appear unclear or ambiguous. Any customer should prefer to clear up details which are not understood fully at the earliest stage rather than discover such difficulties later. A positive decision must be made on such points.

QUESTIONS

1. The various maximum and minimum values given in specifications for times, pressures, temperatures etc. are

   (a) only a set of guidelines to be followed with up to 50% variation accepted
   (b) fixed requirements and must not be varied beyond 5% either way
   (c) limits within which the process must be controlled
   (d) only to be applied when critical components are processed

2. The major problem which can arise when a specification relies heavily on another published document is that

   (a) the published document used as a reference may be changed or even cease to exist so making the specification obsolete also
   (b) anyone operating the specification must obtain and read a second document
   (c) quality auditors must be fully acquainted with the second document
   (d) the second document may issue quality assurance standards which are not the same as the prepared specification

3. Deviations from a specification
   (a) Can be put into operation and the specifying authority informed within 4 weeks
   (b) can only be implemented after receipt of written agreement from the specifying authority
   (c) are never allowed
   (d) can be implemented after verbal discussion with the specifying authority

4. The personnel involved in penetrant processing must

   (a) receive proper formal training in the method
   (b) be checked for visual acuity every 12 months
   (c) be checked for colour vision every 12 months
   (d) all of these

5. When inspecting processed items inspectors

   (a) must take a break of 10 min after 2 hours continuous inspection
   (b) can start inspecting for fluorescent indications in a darkened booth immediately after checking the ambient white light levels
   (c) must avoid fatigue by ensuring that he/she inspects more than one type of component
   (d) must check a replica transfer to ensure that he/she is ready to start inspecting

6. Many specifications for penetrant inspection state the stage of manufacture at which the process takes place. This is often given as
   (a) at any time during manufacture
   (b) after a thorough washing and drying
   (c) after any process which might cause surface defects and before any process which might close them up
   (d) after drying in an oven at 120 °C (248 °F) for 30 min and then allowing them to cool to 40 °C (104 °F)

7. When a penetrant process specification covers the use of detergent (hydrophilic) remover solutions the following should be clearly stated:
   (a) the method of application, the concentration of remover in water and the maximum contact time
   (b) the concentration of remover in water, the maximum temperature and the method of application
   (c) the maximum contact time, the maximum temperature, the method of application and the concentration of remover in water
   (d) the method of application, the maximum temperature and the maximum contact time

8. When aqueous developers are included in a penetrant specification the following points must be covered:
   (a) the concentration of the developer
   (b) the temperature allowed for drying
   (c) the minimum contact time
   (d) all these points

9. When an automatic viewing apparatus is used the specification should
   (a) be written for the specific application(s) involved
   (b) ensure that the viewing conditions using human eyes are replicated as closely as possible
   (c) account for the fact that some human eye inspection is always unavoidable
   (d) limit the number of components inspected per hour

10. Quality assurance standards are
    (a) useful subsidiary documents to help inspectors in their work
    (b) documents which must be kept up to date for the Quality Auditor to see
    (c) essential associated documents which regulate the inspection stage of the process
    (d) normally incorporated into the operating specification

Note: There is only *one* correct answer to questions 1–10. Tick your choice and check it with the correct answers on p. 215

10

Health and safety in penetrant testing

10.1 INTRODUCTION

The need to take account of health and safety is not specific to penetrant processing but is shared by all industrial and even domestic activities. This chapter assesses the known risks to operators while using the materials and ways of dealing with them. Fortunately, the chemicals used in the formulation of penetrant materials tend to be those which present low hazards. Clearly the possibility exists that a chemical which has been in use for some time may display some unpleasant long-term effects.

The continuing monitoring of a wide range of materials over many years offers the opportunity to control such possibilities. It may be preferable to use a raw material of which long experience of low hazard exists than to use a newer chemical which may not be listed in any index of hazardous chemicals yet may with increased use be found to have distinctly unpleasant properties. Where irritant, toxic, flammable or other unpleasant properties are well defined measures can be taken to use them in such a way that operator safety can be reasonably assured. In effect all chemicals should be treated with respect when in use – even water can be considered a hazard in certain circumstances as we can drown in it. Use of chemicals in processing should be controlled by positive decisions based on current data and such decisions should be reviewed regularly to take account of changes in available information.

The control of the labelling of chemical products and the control of safe procedure for workplaces are covered by regulations which have the force of law in many countries. A selection of these, together with standards which cover various protective apparatus, are listed in the appendix to this chapter. Such regulations are under constant review and it is always advisable to check that the current edition is in use before any decision is taken on how to deal with a specific problem.

10.2 HEALTH AND SAFETY ASPECTS OF PENETRANT PROCESSING

This section is divided into two parts, the first dealing with physical aspects of health and safety and the second with chemical aspects. In both cases possible hazards with a wide range of penetrant processes and situations are considered. Anyone planning, designing or operating a penetrant process must assess any health hazard which might arise and take action to avoid foreseeable problems.

10.2.1 Physical safety

(a) Physical injury

The possibility of physical injury to an operator working with penetrant processing depends very much on the equipment used. When the penetrant materials are applied using aerosol spray cans, in which case no other equipment apart from absorbent cloths or paper are involved, physical injuries are normally to the eyes owing to misdirected jets of material or failure to take account of wind directions when working out of doors. Explosion of the can, as a result of exposure to high temperatures, is a third type of physical hazard which can occur. The first two problems can be avoided by using a face shield and the third by ensuring that aerosol cans are not exposed to temperatures higher than those stated by the manufacturer (often 55 °C (130 °F)).

When equipment includes reservoirs of liquid there is the possibility of an operator drowning. Although this may sound unlikely, it has happened. All reservoirs of liquid material must be set out in such a way that accidental or deliberate access to them is difficult and, especially where large tanks of liquid form an installation, nobody should be allowed to work alone on the line. When large numbers of components are processed in a penetrant installation and when large or heavy components are processed, it is normal for some form of handling to be available. Unguarded handling devices are a well-known hazard to operators. There are particular dangers when components are jigged on a special carrier. From time to time one or more will fall off the carrier and there is a natural desire on the part of many operators to rescue and replace them. It is essential that operators are prevented from doing this. This may be by use of physical barriers which cannot be opened while the installation is working or a sensing device which stops the line if anyone should attempt to interfere manually in any way whilst the handling equipment is operating. It is not enough to prevent accidents to operators by instruction alone, even when such instructions are repeated on notices around the equipment.

Conventional spray equipment generally presents few operator hazards apart from the fact that a misdirected jet can cause material to rebound into an operator's face. This physical hazard can be dealt with by using a face shield, as can the hazard of splashes from an immersion process. Electrostatic spraying introduces a number of specific hazards associated with the high voltages used in that process. Manufacturers of such apparatus design it in such a way as to reduce electrical hazard to the operator to a minimum. It is essential that the manufacturer's instructions for installation and maintenance of such apparatus are followed, and in the case of manual use of such equipment the manufacturer's instructions for operation must be followed.

Drying ovens and hot air blasts should not present a serious hazard as far as hot surfaces are concerned because for technical reasons the temperature at which components are dried is normally not above 60 °C (140 °F). Where temperatures above this are used for drying components, as is often the case after pre-cleaning and before penetrant application, the components and carriers should be stored in a cooling area where the possibility of contact with human skin is eliminated.

A genuine hazard on many penetrant installations is the use of the aptly named guillotine type door which is often fitted to tunnel type dryers and storm cabinets for the application of dry powder developers. When guards are not provided around the entrances and exits to such tunnel stations there is a real danger to fingers, hands, arms and even heads in cases of extreme misjudgement. Where such tunnel stations are used it is preferable for the doors to rise to close rather than falling. In either case, it is possible for a sensor to be fitted to the operating mechanism so that if the door is impeded before reaching the closed position it retracts. Such a device might well save some components as well as preventing broken bones, bruises or other injuries.

When electricity is used on a penetrant installation it must be installed and maintained in accordance with good safe practice as covered by the many legal requirements in various countries. This applies to supply to power hoists, motorized roller tracking, fans, heaters, powder storms and lamps, both white light and UVA (black light), inspection booths and wash stations. Particular care must be taken to ensure that the lamps in wash stations are electrically secure as water and electricity do not form a safe mixture.

(b) Dangers of inspection

In the case of lamps there is the considerable problem of heat. Unfortunately it is not possible to separate the various effects of curent electricity and use only the one specifically needed. If electricity is to give light or UVA it will give heat. This is a significant potential hazard when UVA (black light) is used to inspect components after fluorescent penetrant processing. Components are normally inspected in the dark after fluorescent penetrant processing. This is itself a unusual hazard as few industrial processes are carried out in the dark for obvious and good reasons. In addition, the level of UVA (black light) needed to ensure proper inspection means that the source must be a high pressure mercury arc discharge lamp or other high energy source and this involves levels of heat. The bulbs used for such sources and frequently the ballast used to regulate them both become hot, reaching temperatures of well over 100 °C (212 °F) which is sufficient to burn an inadvertent hand touching them. Fortunately the filters which are used in such lamps allow some blue violet light to pass so that they can be seen. A true UVA source (black light) would be a significant hazard to operators, and since all visible light is white light, this fact must be considered when limits of white light emission from UVA sources are specified. Opperators must take care to avoid touching the hot part of lamps.

During recent years UVA Lamps have become available which remain cool to below 70 °C during use even on the Woods glass filter and some newer specifications require that the lamps must operate at such lower temperatures. In this way the hazards associated with UVA Lamps is disappearing.

A further possible hazard exists with sources of ultraviolet light. The sources of the UVA which may be a high pressure mercury discharge arc or a halogen lamp emit energy across a wide range of wavelengths. These produce energy in the UVA area which is needed for inspection following fluorescent penetrant processing and covers the wavelengths between 310 and 380 nm. The mercury arc also emits white light at wavelengths between 380 and 710 nm and in the region of the spectrum known as UVB where wavelengths are below 310 nm. The white light and the UVB are both filtered out optically using a special glass. This is known as a Wood's glass filter or alternatively a Kopp filter. Some UVA lamps use an arrangement where the Wood's glass is incorporated in the bulb. The actual envelope of the bulb is made from Wood's glass and appears dark purple. Other lamps use a mercury lamp with a separate Wood's glass or Kopp filter mounted in front of it. In this case the bulbs have a clear or slightly frosted glass face and a dark purple glass filter is used with them.

UVB contains some wavelengths which are hazardous to life forms – a fact recognized in the use of UVB for sterilization of some medical and other products. Some of the wavelengths in the UVB can cause erythema of the skin, which is better known as sun burn in its mildest form but can be quite painful if exposure is more severe. A further hazard is photokeratitis of the eye which can cause severe problems. None of the activities mentioned above are associated with exposure to UVA and so these hazards are controlled by using a filtered UVA lamp. It is difficult to see how an operator could be exposed to these dangers whilst using a lamp which has the Wood's glass filter integral with the bulb since if the bulbs are broken they stop working. Where the bulb and the filter are separate great care must be taken to ensure that the lamp is *never* switched on without the filter in place, that the filter is not damaged and that it is fitted properly. If any white light can be seen, then UVB is escaping and such a lamp must *not* be used until it is properly repaired. All lamps should be inspected for defects when cold before each period of use.

The problem of UVB causing photokeratitis in the eye must not be confused with an effect of UVA on the liquids in the human eyeball. If anyone looks directly into a UVA lamp, which is a strong source, with unprotected eyes the liquids in the eyes will fluoresce. This is a reversible but sometimes unpleasant experience which can be avoided by using UV-absorbing glass spectacles which may be of plain sodium glass or spectacles which correct eye defects. Do not use photochromic spectacles under UVA as some spectacles of this type become irreversibly tinted upon such exposure. There is also the technical objection that such spectacles reduce the visual efficiency of the inspector.

In recent years concern has been expressed about the potentially harmfull effects of exposure of the skin to UVA. Effects simular to those experienced on exposure to the UVB region have been reported at wavelengths below 340 nm. The exposure which was needed to cause problems was much greater than with UVB with studies indicating ten times the exposure and even higher levels. There appears to be relatively little potential hazard at the levels experienced in inspection booths however despite this and the long and relatively trouble free experience which exists in such circumstances operators and inspectors may wish to guard against any possible hazard by wearing gloves during inspection, PVC is a very good block against UV, or using skin creams which block IV.

A further danger is that cold water on hot glass can lead to the glass cracking and sharp particles of glass being thrown violently over inspectors. Normally inspection booths are dry but sometimes lamps are taken to areas where splashing may occur. Care must be taken to avoid water splashing on to hot bulbs or filters.

10.2.2 Chemical safety

Chemical safety begins with formulation. It is the case with all industrial chemicals that no formulation chemist chooses dangerous raw materials where less hazardous materials are available. However, there are occasions when some level of hazard must be accepted if a job must be done. In such cases sensible precautions must be taken to ensure safe use of the materials. It is obvious that operators cannot take precautions if they are not informed of possible hazards. This is achieved by controlled labelling and by issue of health and safety data sheets for products by the manufacturer or supplier. The format and in some instances the actual wording of parts of such documents are controlled by regulations and law. A list of some of the regulatory documents is given in the appendix.

Some general instructions should be followed whether they are legally required on labels or other documents or not. It is extremely foolish to eat, drink or smoke whilst using chemicals whether in industrial or domestic applications. Transfer of chemicals to the mouth is almost inevitable when eating or drinking while using chemicals. When smoking, the mixture of chemicals taken into the lungs is harmful enough without drawing vapours from the chemical mixture through a cigarette, cigar or pipe and complicating the mixture further. The warning to keep chemicals secure from children is another universal safety requirement. While children should not have access to industrial chemicals at any time, it is as well to remember this obligation when working on a site where children may have access even if such access is the result of trespass or is otherwise unauthorized. This precaution must also be remembered when disposing of chemicals.

The operation of penetrant processing involves, for the most part, materials which present fewer health and safety hazards than do many industrial chemicals.

(a) Penetrants

Many penetrants contain mineral oils, and the hazards associated with these materials are well documented. Modern materials avoid the use of polycyclic aromatic hydrocarbons. It is fortunate that mists of penetrant are not formed in manual operation. The oils used in modern penetrants have quite high boiling points, often above the range 190–290 °C (460–555 °F), as do the final blended products. At normal temperatures, i.e. up to 50 °C (122 °F), the evaporation of penetrant materials is low so that the hazard to operators is minimized. It is still desirable for tanks full of penetrant to be equipped with some form of extraction to ensure a healthy working environment. When penetrant is applied by manual spray good extraction which takes the material away from the operator is essential. When a manual electrostatic spray system is used for penetrant application care must be taken in the design of the spray booth to ensure that swirling of the sprayed material is avoided as this may cause sprayed material to migrate back to the operator. Legal requirements for the rate of extraction for this type of spray must be met. In all cases of conventional spray application extraction should be designed to take air from behind the operator. When spray application of either the conventional or the electrostatic type is incorporated in an automatic penetrant processing line care must be taken to ensure that mists do not escape into the workplace. Fog application cannot be used completely safely except in automatic installations. Just as aerosol cans are likely to suffer abuse from a technical point of view so they present the greatest potential for contaminating the workplace. When applying penetrant from aerosol cans all operators must take very great care to protect their own well-being and that of their colleagues around them who may or may not be involved in penetrant testing. When aerosols are used in a covered workshop care must be taken to ensure that local ventilation is adequate. There are occasions when penetrant testing must be carried out inside a vessel or pipe. In such circumstances the best solution is for the operator to be supplied with an independent air supply so that the local atmosphere is not inhaled. The best form of this arrangement is the enclosed helmet variety which protects the head completely.

Where penetrant is applied by immersion there is the possibility of skin contact if protective gloves and aprons are not worn. When this possibility exists suitable protective gloves, aprons and face masks should be worn. Good practice avoids splashes, but a face mask, or at least goggles, will prevent a nasty experience. Oil-resistant gloves and aprons are usually suitable; however the type should be specified in the penetrant manufacturer's health and safety data sheet.

The normal ranges of penetrants should not be used at temperatures above 55–60 °C (130–140 °F) and this is normally stated. The special penetrants for use at temperatures up to 200 °C do not normally contain hydrocarbon oils, and so the health hazards associated with this group of materials are avoided.

(b) Detergent removers (hydrophilic removers)

The potential for health hazards with these materials must be considered from two viewpoints. The materials are normally used as solutions in water, and these range from 2.5% up to 30% detergent in water according to the particular material which is used. At such concentrations any initial hazard due to the concentrate is reduced to levels similar to that of domestic washing-up liquid. Despite this, it is wise to avoid frequent or prolonged skin contact as all detergents de-fat the skin, dry it and eventually crack it. The skin is a protective layer on the body and removing all the natural oils reduces its protective capability. It is unusual for detergent removers to be applied by any method other than immersion. However, when it is applied by spray or as foam the concentration tends to be less than 2.5%. Water is much more volatile than the detergent concentrate and dangers from mist or spray are minimal.

The second condition in which these removers must be considered is as concentrates from which the working solutions have to be made. The concentrates are very stable and do not evaporate to any significant extent at temperatures below 55–60 °C (130–140 °F) so that inhalation hazards are unlikely. The concentrates do have a strong de-fatting effect on the skin. Skin contact with the concentrates is to be avoided, as it is quite an unpleasant experience. Gloves and aprons must be used when handling the detergent removers in concentrated form.

(c) Emulsifying agents (lipophilic removers)

Emulsifying agents combine the possible hazards of penetrants and the detergent remover concentrates, and so it is probably fortunate that they have been largely replaced by detergent remover solutions. Most emulsifying agents designed for penetrant removal contain mineral oils as diluents to control the speed of emulsification. The actual emulsifying agents are excellent solvents for fats and oils, and so they exert a strong drying effect on the skin. The positive point in their favour is that they generally have very high boiling points and are stable. They do not evaporate to a significant extent at temperatures up to 100 °C (212 °F) so that even when sprayed the tendency to form mist is minimal. The major precautions are to avoid skin contact at all times and to use extraction if the material is sprayed, despite the stability of the product. Always read the manufacturer's health and safety data sheet before using the specific material.

(d) Solvent cleaners

In penetrant testing the term solvent cleaner always refers to a volatile organic solvent. These materials are most widely used in the application of colour contrast penetrant testing. The choice of type of material has always been limited to non flammable and flammable materials and has been further restricted to flammable materials with the elimination of 1, 1, 1, trichloroethane. All of the useful non flammable halogenated hydrocarbon solvents which are not implicated in the depletion of the earth's ozone layer are not acceptable for manual use owing to their unacceptable health and safety characteristics thus leaving only the flammable materials. These include alcohols, esters, and ketones many of which have very attractive health and safety characteristics apart from the fact that they are flammable and form explosives with air. The TWA (time weighted average) figures for many of these materials is high even at 8 hours however at concentrations between 1 and 12% by volume they form explosive mixtures with air. Such levels build up quite quickly so precautions must be taken to allow their safe use. In all instances they must be used in conditions of good ventilation and in the total absence of sources of ignition. This includes circumstances when the air is very dry for climatic conditions which may occur in places where it is very hot and dry and those where it is very cold and dry. In such circumstances electrostatic discharges occur when people shake hands or touch any other conductor and this is enough to start a catastrophic fire.

A significant point is that the hazzards associated with these materials are well known and their management well established. Despite this problems do arise. The author actually witnessed an example of this. The inside of a large vessel had been inspected by a colour contrast penetrant process with a flammable volatile organic solvent and a non aqueous suspension developper which used the same solvent completing the process. About 4 hours later the shop foreman decided to look inside the vessel and, thoughtlessly left a lighted cigarette in his mouth. He was lucky in that the result was a fright and some singed eyebrows and moustache.

All suppliers of all chemical materials have a legal responsibility to warn their customers of any risks and hazards associated with their use. This is achieved by proper labelling and the provision of health and safety data sheets. It is the responsibility of any specifying authority to repeat this information and it is the responsibility of anyone using such materials to familiarise themselves with their proper use to ensure their own and their colleaque's and any other bystander's safety.

(e) Dry powder developers

Dry powder developers consist of very fine particles ranging in size from submicron to around 12 μm. Such materials are a nuisance if it gets into the air in any quantity. They are physical irritants particularly to the eyes, nose and throat. When used by storm application the developer must be kept in an air-tight box which cannot be opened until the powder has settled. Since normal settling takes a very long time, it is usual for powder storm cabinets to be equipped with extraction apparatus to clear the volume of air before the developer station can be opened. Where dry powder developer is applied by electrostatic spray or floc gun in a manual operation good extraction is essential and it is advisable for the operator to wear a nose and mouth filter to prevent inhalation of particulate matter and eye protection in case of mishaps. When dry powder is applied by dusting on or by using a rubber pear puffer the quantity of developer available is small and the risk of irritation is reduced. Despite this, due care must be taken to avoid creating nuisance.

After the identification of asbestosis as an industrial disease, concern arose about dry powder developers since some formulations contain talc and many sources of talc are contaminated with asbestos fibres. Fortunately there are a number of sources of talc in the world which are not contaminated with asbestos and these have been used in penetrant developers for a very long time since the presence of asbestos even as a contaminant in developers causes technical difficulties.

(f) Non-aqueous suspension developers

Non-aqueous suspension developers are also known as 'liquid spray developers'. These materials are mixtures of developer powders suspended in volatile organic solvents. In application the problems of free powder in the air do not arise as the powder is thoroughly wetted and incorporated in the suspension. However, some of these materials become very dusty after application and care must be taken when they are cleaned off. Other formulations leave layers of developer which are essentially water soluble and can be removed using warm water, while others incorporate materials which hold the developer powders together even when dry. While such materials present a problem for cleaning components they present less of a health hazard.

The major hazards with this type of developer are associated with the solvent carrier. These are very similar to those outlined in section 10.2.2 (a). The hazards are continuous, particulary since for technical reasons these materials must be applied by spraying. The choise of materials is restricted to flammable materials such as alcohols, esters, and ketones. All suitable materials of these types have flash points below 20 °C (68 °F) and are classified as highly flammable. The third type of solvent which is technically suitable is the family of CFCs which are environmentally unacceptable because of their potential effects on the earth's atmosphere (the ozon layer). Current research on other solvents may give volatile solvents which are non-toxic, non-flammable and do not have properties which suggest that they threaten the environment.

In using any form of organic solvent in any way care must be taken to avoid contact with eyes and skin. It is also essential to avoid taking these materials into the mouth. If, by accident, such material does get into the mouth, wash out immediately with large quantities of water and avoid swallowing. Then obtain medical treatment directly afterwards.

Since these developers are applied by spray, frequently from aerosol cans, proper precautions must be taken. Ensure that the spray nozzle of any aerosol can is pointing away from the user or anyone else. When working indoors ensure that good ventilation is available. If the material must be used inside a pipe or closed vessel, the operator should have an independent air supply or, at least, a suitable breathing mask. Under such circumstances the operator's skin should also be protected. When aerosols or other spray apparatus are used on site it is a wise precaution to check the direction of the wind and avoid the possibility of sprayed materials being blown back over the operator. Since there are occasions when the operator has no choice as to the direction of spraying, e.g. application to large fixed structures or large heavy fabrications, and there is the possibility that the sprayed material will swirl and return over the operation, in many outdoor situations it is necessary to wear a face mask or goggles to protect the eyes when applying non-aqueous suspension developers.

Despite the potential hazard of using these materials reports of accidents to operators have been very rare over the past years. In view of the quantities used it is difficult to put this down simply to good luck. It appears that operators are reading the warning labels on the packages and heeding them.

(g) Aqueous developers

It is possible to consider the health and safety aspects of both types of aqueous developer under one heading since both the water-soluble and water-suspendable developers present a similar type and level of safety consideration. Both types of product are normally supplied as powder concentrates for mixing in water at levels between 30 and 300 g/l according to individual product. As powders these materials are fairly inert. They contain surfactants and wetting agents which have a detergent action on the skin and remove the natural oils, leading to drying and cracking. The effect of the powders in the eye is to cause severe stinging through the detergent action and this must be avoided. It is possible although unlikely, that the powders could be transferred from hand to eye while mixing the working solution. For this reason alone it is a good idea to wear rubber or PVC gloves while mixing aqueous developers. This precaution will also protect the hands from the drying effects of the materials.

Once the working solution has been made up the detergent effect on the skin and eyes is reduced. However, if hand contact is likely to be frequent or prolonged, rubber or PVC gloves must be worn and eye protection is needed if splashes are likely or if the developer is applied by spraying.

10.2.3 Aerosols

A considerable proportion of colour contrast penetrant testing is carried out using aerosols. This often takes place in an open workshop or outdoors on site. When the inspection is performed in a workshop good ventilation is essential. If the general ventilation in the shop is not adequate local extraction is required. Outdoor work carries its own hazards, of which one is wind. If the wind is strong great care is needed when aerosols are used. It is not simply a question of the wind blowing sprayed material the wrong way, but also the problem of swirling. Suitable gloves and a face mask can prevent unpleasant experiences in these circumstances.

A problem which can arise either in a workshop or on site is that the inside of a large pipe or vessel may need inspection and this requires the operator to enter the structure. In such circumstances the only material available is often in aerosol form. Under such circumstances an independent air supply is essential and the entire operator should be protected. The operator must also be accompanied by someone who can raise the alarm if a mishap occurs.

Aerosol packages present their own hazards as they are under pressure and require a propellant which may have its own particular hazardous property.

The fact that aerosol cans are under pressure means that they can become dangerous if punctured or heated. The most likely cause of puncturing is physical damage on site. Aerosol cans are sometimes carelessly placed and run over by vehicles. Even when simply lying on the ground they are an obvious danger to passers-by or even operators themselves who might trip on them. After being run over by passing vehicles the aerosol cans may be totally flattened, in which case they are no longer a hazard or they may survive in various states. After being run over they almost always become weakened and the danger of sudden release of the contents is possible, if not likely. This author has seen the results of someone investigating aerosols damaged in this way on several occasions. Fortunately the results have been comic rather than serious – on each occasion so far with unfortunate red trousers rather than the potentially tragic red face. There is only one way to deal with an aerosol can after it has become damaged – dispose of it. If it must be picked up, do so with a remote device such as a mechanical grab on a stick, and then confine it in a box or cover it with wet earth or sand sufficient to contain the release of material. The best way of dealing with this danger is of course proper management of the cans while in use and ensuring that they cannot be run over by vehicles or otherwise crushed. The problem of heating volatile material which is confined in a fairly strong can comes when the pressure inside becomes greater than the strength of the can. The can fails and the excess pressure is released suddenly. It is very rare for a can to shatter – it usually splits at a weld – but the energy released is very significant. The actual amount released depends on the actual contents of the can but it is always enough to cause damage. No aerosol can should ever be heated above 55 °C (130 °F).

The choice of propellant for aerosols is at present limited. For many years CFCs have been used in aerosol chemicals for industry including penetrant materials. The expense of these propellants was offset by the fact that they present low levels of health hazard to the user with typical TWA values of 1000 ppm, are non-flammable and work at low pressure. Unfortunately the most useful members of this group of chemicals are strongly suspected of causing depletion of the ozone layer in the earth's atmosphere and have been largely phased out. At the time of writing the related fluorocarbons which do not contain chlorine and do not pose the same environmental problem have not become available for industrial use. There is a propellant 22 available which offers a partial solution and offers a reduced environmental threat. This leaves the choice of a compressed gas such as nitrogen or carbon dioxide or of the hydrocarbons propane or butane or various mixtures of these two propellants.

The compressed gases are not hazardous in themselves but they do lead to much higher pressure in the cans, and so any explosion occurring will be worse than that caused by puncture of an aerosol can using some other propellants. The use of propane, butane or mixtures of these propellants involves lower pressures than carbon dioxide and some other alternatives but the flammability of these materials is such as to raise concern about their use when electric sparks or other sources of ignition may be present. In enclosed volumes of air there may be problems of concentration build-up unless ventilation is efficient.

Where aerosol cans of penetrant materials are responsibly and sensibly used hazards are controlled. Fortunately, safe working practice and respect for the possible dangers due to misuse is normally found.

10.2.4 Documentation for health and safety

Health and safety documentation for industrial chemicals can be divided into two sections: labelling, and health and safety data sheets.

(a) Labels

The major use of labelling is to inform people who are involved with storage and carriage. In most countries there are regulations governing what can be carried by road, and regulations governing air and sea transport are observed internationally. There is a legal requirement for manufacturers to declare any hazardous chemicals. In this declaration is included the requirement to declare the risk and the safety precautions which must be taken. Examples of the symbols used and the internationally approved risk and safety phrases are given in Fig. 10.1. The carrier must check the data given against local requirements as these may vary. An example of variation is that a liquid is designated combustible in Europe if its flash point, as measured by the Pensky Martens closed-cup method, is below 100 °C (212 °F); in the United States this classification is applied to materials with a flash point measured by the same method below 200 °F which is just above 93 °C. While differences between requirements for different countries are generally small, they have legal force and must therefore be followed carefully. Similarly, storage requirements have legal force and show minor variation from country to country.

The sources of data and classification of chemicals are various and some are listed in the appendix. The characteristics of many chemicals are well

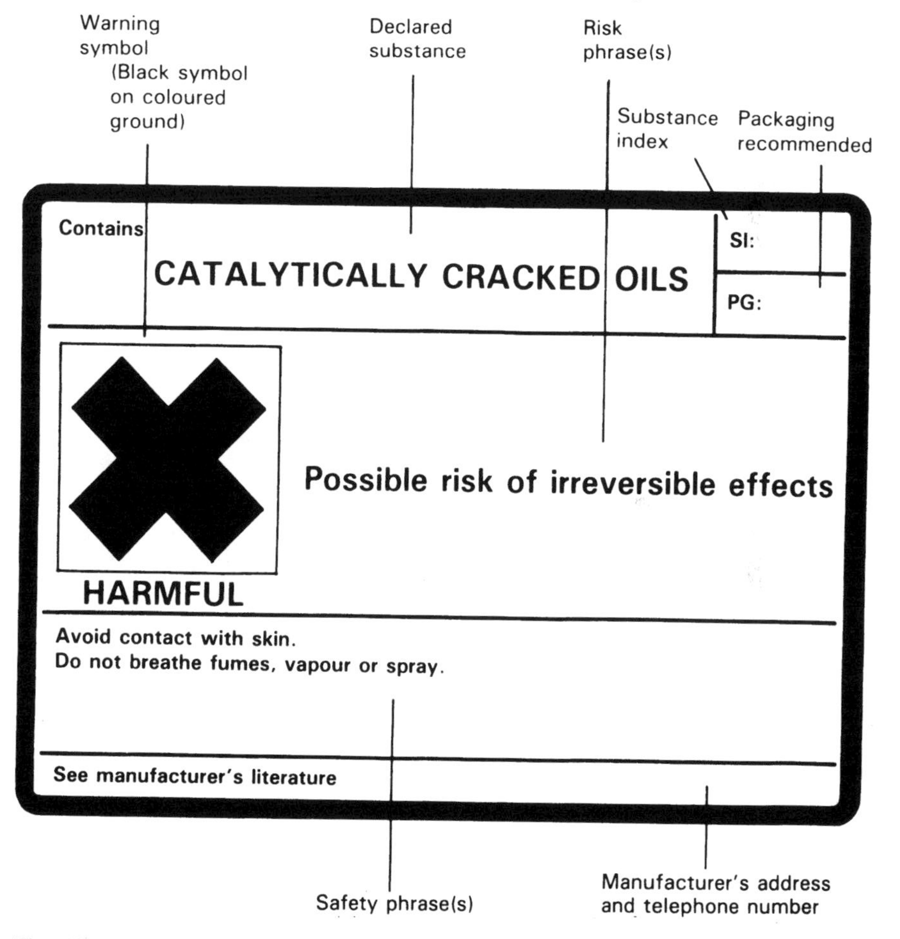

Fig. 10.1 Typical safety symbols for labelling industrial chemicals.

documented and are listed by various agencies. The list of chemicals used in formulations of penetrant materials and other specialized industrial chemical products has grown significantly over the past 50 years and some are quite new.

Few chemicals are used solely for penetrant materials and the great wealth of research into the acceptability of the raw materials is available. When penetrant materials are formulated every effort is made to avoid dangerous chemicals. However, undesirable properties of some raw materials may become apparent after the formulation has been established. In such cases the materials involved may not appear in lists of hazardous materials, but the manufacturer has a legal requirement to revise the label and declare the presence of any material suspected of any hazard whether listed or not. Clearly, if research showed a very serious potential hazard which had previously been unsuspected and which would make safe operation of the process difficult, the penetrant material would be reformulated if possible or replaced by another product.

Fortunately, most penetrant materials carry relatively low risk and are often classified as 'no significant hazard'. In the cases where the materials are harmful or flammable the hazards are well documented and can be managed by proper use. Manufacturers of materials can label the containers but this is useless if nobody reads the labels. The risk and safety phrases which are now defined by law are very direct and simple to understand.

(b) Health and safety data sheets

It is a legal requirement in many countries to provide health and safety data sheets for industrial chemical products. In some countries standard forms are used. Examples include Germany, where forms according to the standard DIN 52900 are used, and the United States, where forms based on the US Department of Labour, Occupational Safety Health Administration form OMB 1218-0072 are used. Sweden and Norway also have their own forms for health and safety purposes. At present the UK requirement is for the supply of data sheets which meet the requirements of the 'Health and Safety At Work Act 1974' and amendments to this act (Table 10.1). At present the act requires that the health and safety data sheets must indicate the following:

- product name
- supplier
- warning of any hazard
- advice for safe handling and storage
- emergency and first aid procedures
- fire protection data
- a full description including the identify of any potentially hazardous raw materials
- data on the environmental impact of the material which includes comment on handling, spillage and disposal of used or unwanted material
- date when the document was prepared

Table 10.1 Information which should appear on a material's health and safety data sheet

Product name:	
Application:	
Warning statement	If 'no significant hazard' this should be stated
Hazardous materials	Any hazardous materials included in the product must be identified by their chemical name, CAS or other registry number, their concentration stated and where appropriate their TWA value
Safe handling information	Storage
	Actual use – including protection from contact, need for local exhaust etc.
Emergency first aid procedures	Eyes
	Skin
	Swallowing (ingestion)
	Breathing into the lungs (inhalation)
	Aspiration (introduction into the lungs)
Fire protection data	Flash point
	Suitable extinguishing media
	Any special fire hazard
Description	Physical state
	Colour
	Smell
Protection of the environment	Environmental impact
	How to deal with spillage
	Waste disposal method
Physical data	Boiling point/melting point
	Specific gravity/bulk density
	Vapour pressure
	pH of concentrate
	pH of working solution
Chemical reactivity	Polymerization
	Conditions to avoid
Manufacturer/supplier	Name
	Address
	Telephone number
	Telex number
	Facsimile number
Date of preparation of data sheet	

In practice two problems may occur. One is that the health and safety data sheet may not reach the operators despite having been supplied to the company involved. In the United Kingdom these documents may be sent to the company secretary as that officer has legal responsiblity for such matters. Where safety officers or occupational health officers are known to exist in a company suppliers will send the documents to them. In other countries the supplier will send copies of these documents to safety officers or occupational health officers where known, or alternatively to the company officer who is legally responsible for ensuring that the data have been provided. The other problem is to ensure that the health and safety data available to the user are up to date. Some suppliers have procedures to ensure that revised sheets are sent to customers.

One excellent precaution which is sometimes used on fixed installations is for a label to be prepared for each station which gives the product name and a brief description of the formulation (Table 10.2). These labels are best prepared in duplicate, one fixed and the other removable. If an operator does suffer extensive skin contact or severe eye contact, or swallows some material, the removable label is taken to the first aid and medical staff so that they can give appropriate treatment without delay.

Table 10.2 A selection of risk and safety phrases used on the labels of industrial chemicals

Risk phrases	*Safety phrases*
Harmful if swallowed	Keep out of reach of children
Harmful by inhalation	Keep away from heat
Flammable	When using do not eat or smoke
Highly flammable	Do not empty into drains

10.3 STORAGE AND HANDLING

Good normal storage in dry conditions away from direct sunlight temperatures above 55 °C (130 °F) is all that is needed for most penetrant materials. Container volumes typically 5, 25 and 200 l for liquids and 1, 5 and 25 kg for powder materials. The exceptions are those materials with flash points below 55 °C (130 °F), which are designated flammable, and below 21 °C (70 °F), which are designated highly flammable, and storage of these materials is governed by law.

Aerosol cans present their own special problems as they contain material under pressure. These packages must not be stored at temperatures above 55 °C (130 °F). The commonest cause of problems with aerosol storage is the result of stacking too many packages on top of one another. Eventually the weight becomes too great for the cartons which start to collapse, and eventually one or several actuators are triggered and a quantity of material becomes spoiled. If a volatile solvent is involved, any health hazard associated with the material must be respected.

Similarly, handling of penetrant materials is for the most part not seriously hazardous. When penetrant is first poured there may be some smell. However, this normally decreases after a few hours. When making working solutions of the detergent products (hydrophilic removers) the precautions outlined in section 10.2.2 must be taken, and when powder storm cabinets are loaded with dry powder developer care must be taken to avoid escape of powder to the working area.

Fortunately, the materials which contain volatile solvents are not normally transferred from the original containers until just before use. On the occasions when these materials are loaded into spray apparatus, all the precautions associated with using the material must be observed.

10.4 CONCLUSION

It is fortunate that much penetrant testing involves the use of chemicals which present, at worst, low hazard. The two product types which contain volatile organic solvents present increased hazards, but this is well known and documented and has been managed successfully.

All chemicals should be treated with respect, and skin and eye contact must always be avoided wherever possible. Eating, smoking or drinking while working with penetrant materials must be avoided. If clothing accidentally becomes soaked with penetrant it must be removed, and eye protection should be worn if splashes or sprayed droplets are likely. Sensible working practices with penetrant materials will avoid mishaps.

APPENDIX

Regulations, Standards and other Sources of Data Concerning Health and Safety Aspects of Penetrant Testing

1. *Health and Safety at Work Act 1974 and Amendments*, HMSO, London.
2. Health and Safety Commission, *Classification and Labelling of Substances Dangerous for Supply, Approved Code of Practice*, HMSO, London, 1988.
3. *Health and Safety Guidance Note EH40, Occupational Exposure Limits*, HMSO, London.
4. *Electricity Supply Industry Standard 98–21. Magnetic Particle and Dye Penetrant Testing Materials – Chemical Compatibility and Health Hazards*, National Power Division – CEGB, Stockport, 1989.
5. *SI 1657, The Control of Substances Hazardous to Health Regulations*, HMSO, London.
6. National Radiological Protection Board, *Protection Against Ultraviolet Radiation in the Workplace*, HMSO, London.
7. *Criteria for a Recommended Standard, Occupational Exposure to Ultraviolet Radiation*, United States Department of Health, Education and Welfare, Public Health Service, N.I.O.S.H., US Government Printing Office Stock No. 1733–000–12, 1972.
8. N.I. Sax and R.J. Lewis, *Dangerous Properties of Industrial Materials*, 7th edn, Van Nostrand Reinhold, New York, 1989.

9. *British Standard BS 1651, Industrial Gloves*, British Standards Institute, London.
10. *British Standard BS 2092, Eye Protectors for Industrial and Non-industrial Use*, British Standards Institute, London.
11. *British Standard BS 5345, Code of Practice for Selection, Installation and Maintenance of Electrical Apparatus*, British Standards Institute, London.
12. *British Standard BS 5423, Portable Fire Extinguishers*, British Standards Institute, London.
13. *British Standard BS 5750 (ISO9000) Quality Systems*, British Standards Institute, London.

QUESTIONS

1. Most penetrant chemicals are fairly safe to use if good working practices are observed. However, first aid and medical attention are needed if

   (a) the material is accidentally swallowed
   (b) the material gets into an operator's (or other person's) eyes
   (c) the operator or people nearby feel faint, sick or sleepy
   (d) any of these occurrences need rapid attention

2. When working on an automatic penetrant installation what action should be taken when a component falls from a jig or fixture into a tank?

   (a) stop the installation and retrieve the component to avoid possible damage to the installation and other components.
   (b) open the station at which the component fell and retrieve the component.
   (c) wait until the station at which the component fell is free and then retrieve the component without stopping the line.
   (d) leave the fallen component where it is and inform the supervisor.

3. The UVA (black light) lamps used for inspection of components after processing with fluorescent penetrant processes give off a visible blue light. This is helpful from a safety point of view because

   (a) it is possible to see the outline of components so that they are less likely to be dropped
   (b) an invisible UVA (black light) lamp would be a hazard as it becomes hot in operation and could easily be touched accidentally in true darkness
   (c) the blue light makes dark adaptation of inspectors' eyes occur quickly
   (d) the workbench can be seen easily and parts are less likely to be knocked against each other

4. Some UVA (black light) lamps have separate Wood's glass filters. If the filter becomes cracked it must not be used until a replacement is fitted. Why?
5. Modern penetrants and removers are relatively acceptable as industrial chemicals. The commonest health problem with these is

   (a) inhalation of vapours
   (b) accidental fires
   (c) staining, drying and cracking of the skin
   (d) storage of material in drums

6. Aerosol packages of penetrant materials have a certain specific hazard:

   (a) droplets of sprayed materials are a significant nuisance
   (b) spraying materials from too close a distance can cause 'bounce back' of material to the operator
   (c) they must be stored upside down to ensure that they spray
   (d) all aerosols are effectively a compressed gas and are hazardous if they burst

7. When working in a closed vessel or pipe an operator must

   (a) be equipped with an independent supply of clean air
   (b) use materials which are not packed in aerosols
   (c) avoid using volatile organic solvents
   (d) avoid using colour contrast penetrants

8. Labels on tins of penetrant chemicals must

   (a) identify the product and the manufacturer
   (b) carry the manufacturer's name and telephone number
   (c) identify the product, the manufacturer and any appropriate hazard warning
   (d) carry a hazard warning label

9. Health and safety data sheets are

   (a) a legal obligation on the manufacturer
   (b) a sensible provision by the manager
   (c) a helpful guide to the NDT supervisor
   (d) not necessary if the product offers no significant hazard

Note: There is only *one* correct answer to questions 1–3 and 5–9. Tick your choice and check it with the correct answers on p. 215

11

Management of penetrant waste and environmental considerations

11.1 INTRODUCTION

Just as care must be taken to ensure that operators are not exposed to hazard from use of industrial chemicals, so the environment must be protected. Concern for this factor starts at the stage of formulation and, wherever possible, materials which pose the least environmental threat or which can be dealt with effectively are chosen. Concern for the environment is not new. In the eighteenth century concern was expressed over pollution of the air and rivers from both industrial and domestic sources. During the nineteenth century the original Alkali Act 1863 was passed as a result of the fumes from certain chemical works. This act has been amended, strengthened and extended as the years have passed. Important legislation for control of discharge of effluent into water dates from 1923 with the Salmon and Freshwater Fisheries Act. In 1936 and 1937 Public Health Acts were introduced, and the post-war years have seen the implementation of a number of new measures including the Control of Pollution Act 1974 and the Water Act of 1989. In other countries there has been and remains a commitment to reducing pollution and control of industrial effluent. The United States has its Environmental Protection Agency which exercises control there. There is great international concern on this subject which is strongly supported by public opinion.

Unfortunately, despite various legislation around the world and the significant amount of effort to define harmful effluent and recommend methods of dealing with some materials, the position is not always clear. Expert opinion may be contradictory, with different authorities interpreting data in different ways. It is to be hoped that the intense efforts currently devoted to the problems of managing industrial waste will lead to a clear, practical and successful programme for control of this material.

The considerable interest in the potential and real problems caused by industrial effluent has introduced into widespread use words which were previously unknown outside specialized circles. These include 'ecology', 'environment' (in the scientific sense) and 'biodegradable' among others. The word 'ecology', once almost unheard outside biological science departments of universities and other centres of academic learning, has now become applied to political parties and ideas with, it must be said, a loss of precision in its meaning. The word 'biodegradable' has become very important in the context of industrial effluent, including the wash waters from penetrant processing. It is necessary to discuss what this word means.

11.2 BIODEGRADABILITY

The author attended a lecture during the early years of this word's transition from specialized meaning where the speaker used the word several times. One listener asked what was meant by a material's being biodegradable and received the reply 'it rots'. In the broadest sense this is quite correct, but it is worthwhile going a little further. To what extent does it rot? Some chemicals and materials do not rot or biodegrade under normal circumstances at all. However, very few chemicals or materials are 100% biodegradable and the question of how long it takes for the degradation to take place must be answered. A further consideration must be the nature of the environment into which the material is released. From time to time archaeologists and historians become very excited about animal bodies and sometimes human remains which have been preserved in peat bogs or silt for hundreds or even thousands of years. Here material which is normally readily biodegradable to a very significant extent simply has not degraded. Both the rate and extent of degradation of any material released to an environment depends not only on its intrinsic nature but also on at least the ten following factors in the local environment:

1. the temperature and changes of temperature;
2. the dissolved oxygen concentration in water in the environment;
3. the acidity or alkalinity of the environment;
4. the concentration of salts in the environment;
5. the amount of material released into the environment;
6. the number of micro-organisms present in the environment which can cause degradation;
7. the number of other micro-organisms present;
8. the time of exposure;
9. the concentration of nutrients in the environment other than the material released;
10. the concentration of trace elements available.

Another point which must be taken into account is: What are the products of degradation? Are they noxious, persistent or both?

It is clear that, despite some advertising copy, the term biodegradable cannot be used carelessly. It is not a property of materials which is defined clearly as is viscosity for example. It cannot even be defined as clearly as flash point, which varies according to the method chosen for its measurement. Certainly the indications of the biodegradability of products vary according to the method of measurement. However, there are the added complications that both the extent to which it degrades and the time taken to reach the effective end of degradation must be taken into account.

It is, in effect, meaningless to say simply that a material is biodegradable. The methods used to assess biodegradability can only give estimates of what will happen when the material is released. The rates of chemical reactions involved in degradation, such as hydrolysis, photolysis and free-radical oxidation, can be calculated and measured easily enough in the laboratory. The process becomes very much more complicated when the rate and extent of the reactions are influenced by the state of the natural waters. The type and concentrations of microbes, the levels of nutrients already available, acidity and many other properties of the water have an influence on what happens in a practical situation.

Tests designed to assess the biodegradability of a material fall into two classes; static tests and through-flow tests. When static tests are used, the concentration of the test material is measured against time.When through-flow tests are used a constant flow of test material is fed to a reaction vessel where the total concentration is maintained at a constant value. The proportion of degradation is calculated from the difference between the concentrations at the inlet and the outflow.

Table 11.1 lists four types of test method:

1. static tests in shaken flasks
2. static tests in percolating filters
3. tests with activated sludge
4. tests with continuous cultures

There are very marked differences in the favoured choice of method from country to country and even between authorities in the same country. The method or methods selected depend on a number of considerations including the properties of the material, the quantity of biological material present, the characteristics of the bacteria used and the length of time allowed for the test.

11.3 OXIDATION

The literature on oxidation and the oxidizability of effluent containing waste comprising organic materials mixed with water has its own jargon. A brief description of five of the more commonly found terms is given so that the difficulties in actually measuring some of the properties and the difficulty in assessing the relative importance of each for individual effluent materials can be appreciated. The level of biodegradability of a material is measured

Table 11.1 Methods for the determination of biodegradability data

	Static systems		Flow-through systems	
Aspects	Die away tests in flasks	Die away tests in percolating filters	Activated sludge systems	Chemostats
1. Comparison with environmental conditions	Discharge in river	None	Aerobic purification plant	Natural waters
2. Mechanisms of adaptation	No special mechanism	Growth and dying	Growth not specially related to the substrate	Sufficiently fast growth; other micro-organisms are washed from the system
3. Duration of test	4–8 weeks	First substrate, 9 weeks; subsequently 1 week	6–12 weeks	6 weeks
4. Properties of substrate	No restrictions	Soluble in the medium; non-volatile	Soluble in the medium; non-volatile	No solids
5. Media	No restrictions	No restrictions	Artificial sewage	No restrictions
6. Analysis	Measurement of mineralization Specific analysis	Measurement of mineralization Specific analysis	Specific analysis	Measurement of mineralization Specific analysis Bioassays
7. Equipment	Laboratory glassware shaker	Glass columns and pumps	Perspex or glass system and accurate metering pump	Chemostat accurate metering pump, ph controls
8. Oxygen availability	Aerobic and anaerobic	Aerobic	Aerobic	Aerobic and anaerobic
9 Temperature	25 °C or the temperature of the environment under investigation	20–25 °C	20–25 °C	25 °C or temperature of environment under investigation

by the ease with which it can be oxidized. In this context a range of terms are used which have specific meanings.

11.3.1 Biological oxygen demand

When water which is contaminated with organic material is discharged into lakes and rivers purification takes place by biological action. Biochemical oxidation is carried out by micro-organisms, including bacteria, which use the contaminant as a source of energy. The respiration of these living organisms is supported by oxygen dissolved in the water. Tests for biological oxygen demand (BOD) attempt to produce a model of the complex processes which might occur in natural water. The 5 day version of this test (BOD5) has been adopted in various forms in the full knowledge that it is only a guideline to the time and extent to which effluent becomes oxidized in real situations. In many cases the 5 days of the BOD5 tests have been too short for sensible results to be obtained and longer times have been used. The results of this type of test allow the effect of an effluent on the oxygen resources of the water which receives it to be estimated. The test is a measure of the oxygen consumed by the micro-organisms present while they act on the organic material in the effluent. It normally involves several dilutions in order to find the flasks in which the final concentration of effluent is around 50% of the original.

BOD tests do not give absolute repeatable measurements. Studies showing standard deviations greater than 20% are acceptable and there is no procedure for the determination of the accuracy of these tests.

11.3.2 Chemical oxygen demand and permanganate value

Chemical oxygen demand (COD) and permanganate value (PV) are used to measure the total chemically oxidizable material in a sample. PV measures a specific method of oxidation using permanganate ions; other oxidizing agents such as acid chromate VI are also used. False results for this type of test can be caused by interference from other chemicals such as ionic halides and this can be compensated for by complexing halide ions with a mercury salt before oxidation takes place. COD is not a measure solely of the organic material present but indicates all material present which can be oxidized by these reagents.

11.3.3 Total oxygen demand

In order to measure total oxygen demand (TOD) complete combustion of all oxidizable material in the sample must be achieved. This is outlined in the following chemical reactions where the final reaction products must reach their highest oxidation state.

$C_n H_{2m} + (n + m/2)\ 0_2 \rightarrow nCO_2 + mH_2O$
$2H_2 + O_2 \rightarrow H_2O$
$2N^{3-} + 3O_2 \rightarrow 2NO_3^-$
$S^{2-} + 2O_2 \rightarrow SO_4^{2-}$
$2SO_3^{2-} + O_2 \rightarrow SO_4^{2-}$

11.3.4 Total organic content

The measure of total organic content (TOC) is somewhat specific. A very small sample of aqueous effluent is injected into a catalytic combustion chamber at 950 °C. The oxidation which takes place leads to production of CO_2 which is measured and gives an indication of the amount of carbon present in the initial sample.

11.3.5 Theoretical oxygen demand

The theoretical oxygen demand (Th OD) is the amount of oxygen needed to oxidize hydrocarbons to CO_2 and water according to the following chemical equation.

$$C_nH_{2m} + (n + m/_2)\ 0_2 \rightarrow n\ CO_2 + m\ H_2O$$

The actual theoretical oxygen demand (Th OD) from this chemical reaction is

$$\frac{8(2n - m)}{6n - m}\ g$$

When the organic molecules of the effluent contain other elements such as nitrogen, sulphur or phosphorus, Th OD must take account of these.

11.3.6 Summary of oxidation data

Some years ago the ratio of the measured BOD to the COD expressed as a percentage was widely used as a guide to the biodegradability of a material. This is no longer generally accepted. The fact that the BOD is an inexact measure, and even the COD, TOD and Th OD are open to serious lack of precision because of the presence of other materials, has led to a more cautious interpretation of these values.

It must also be remembered that the various oxidation tests can only be performed on materials which mix with water unless special measures are taken so that pure hydrocarbons and other chemicals which do not mix with water cannot be assessed in this way. From the point of view of penetrant processing this means that only emulsified rinsings can be assessed in this way.

11.4 SOURCES OF EFFLUENT FROM PENETRANT PROCESSING

The type of effluent produced by penetrant processing depends on the type of process in use. The penetrant may or may not be water washable. When the penetrant is water washable it will mix with water to a significant extent or even in all proportions. When the penetrant is not water washable it does not mix with water. The effluent from the process is produced during the penetrant removal stage.

11.4.1 Effluent from the water-washable penetrant process

Water-washable penetrant effluent from manually operated installations contains nominally 1000 ppm of penetration water from the rinse stage. Automatic penetrant processing installations produce effluent water with a much lower concentration of penetrant normally ranging from 100 to 200 ppm. The colour, either red or fluorescent yellow-green, is easily seen even at these low concentrations. Many water-washable penetrants contain hydrocarbon mixed with dyes, emulsifying agents and other surface active chemicals. Some products contain no hydrocarbons at all. The claims that such materials are necessarily more readily biodegradable and so more acceptable from an environmental point of view are difficult to support on the basis of present scientific evidence and should be viewed with caution.

11.4.2 Effluent from the post-removable penetrant process

The main difference between post-removable penetrant and water-washable penetrant is that the latter does not mix with water and another chemical is needed to remove the surface excess. Three methods are used commonly to remove the surface excess:

1. a detergent solution (hydrophilic process)
2. an emulsifying agent (lipophilic process)
3. a volatile organic solvent (solvent removal)

(a) Effluent from the hydrophilic process

The hydrophilic penetrant process involves two removal steps which create aqueous effluent. The first is the pre-rinse step which is used to drive off most of the surface excess penetrant with a water or air–water spray depending on the controlling specification or usual practice. The effluent mixture from this procedure is a two-phase mixture of water and penetrant without emulsification. The water and the penetrant can be separated sometimes simply by allowing the mixture to stand undisturbed for some time. The concentration of penetrant in such mixtures is similar to that obtained from the water-washable penetrant procedures, i.e. 1000 ppm from manually operated installations and 100–200 ppm from automatic installations.

After the pre-rinse stage the components are immersed in the detergent solution (hydrophilic remover) and then washed again with water. The concentration of the detergent solution in water ranges from 5% to 30% depending on the formulation of the penetrant and the remover. The second rinse or wash produces a second aqueous effluent which is quite distinct from the effluent from the pre-rinse. This effluent is a single phase and will not separate on standing. The concentration of contaminants is very low: 50–300 ppm from manual installations and 5–30 ppm from automatic installations. The contaminants consist of a very small amount of penetrant together with non-ionic surfactants and detergents.

(b) Effluent from the lipophilic process

Effluent from the lipophilic penetrant process is essentially similar to that from the water-washable penetrant process. The concentration of emulsifying agent is higher than it is in effluent from the water-washable process because of the characteristics of the method: 200–3000 ppm in effluent from manual installations and 200–600 ppm in effluent from automatic installations. The reason for these differences is that, in using this type of penetrant process, the penetrant is applied and allowed to drain, and after the completion of penetrant contact time the emulsifying agent is applied over the penetrant. At the end of the emulsification time, the activity is stopped using water. More emulsifying agent than penetrant remains on the tested surfaces and emulsifying agents generally have high viscosities compared with penetrants. Emulsifying agents have viscosities ranging from as little as 25 to over 100 cSt when measured at 40 °C (105 °F).

The process itself and the characteristics of the emulsifying agents account for the difference in the mixture and the concentration of material in the effluent water. Qualitatively, the contamination is similar to that in the effluent from the water-washable processes.

(c) Effluent from the solvent removal processes

The solvent-removal penetrant process is rarely applied automatically or semi-automatically in a vapour phase procedure. Such a process ensures maximum recovery of the solvent and so loss of material to the air is minimized. In the manual procedure the solvent is allowed to evaporate and is lost to the atmosphere. At no time should organic solvent for removal of surface excess penetrant be mixed with effluent waters.

The organic solvents used for penetrant removal include non-flammable halogenated hydrocarbons such as 1,1,1-trichloroethane and various flammable solvents such as low boiling hydrocarbons, alcohols, esters and ketones. The volume of such materials released to the atmosphere during penetrant processing is small compared with the overall total of release of solvent vapours from domestic and industrial sources. However, it is possible that some restrictions on the release of some of these materials will be imposed in the future. At present the balance between the safety of non-flammable solvents as far as the operator is concerned and any suspected environmental threat must be considered very carefully. The future may offer non-flammable volatile solvents which do not have the cloud of suspicion of environmental threat over them.

11.4.3 Effluent from the use of developers

Only one type of developer produces effluent in quantities which need serious consideration. The small amount of dust which escapes from a well-managed penetrant installation should not be a nuisance, let alone an environmental hazard. The dust used is often a mixture of naturally occurring minerals, sometimes mixed with a carbohydrate, so that even a large escape would not be considered an effluent problem. Only the non-aqueous suspension developers can be considered to cause effluent. These materials are carried by volatile organic solvents which evaporate after application. These are of the same chemical type as those used by solvent removers, and the environmental concerns over these materials are outlined in 11.4.2 (c).

11.5 DEALING WITH EFFLUENT FROM PENETRANT PROCESSING

A number of methods for dealing with the effluent from penetrant processing are available. The effluent from the water-washable penetrant process, from the second wash or rinse from the post-removable hydrophilic process and from the post-removable lipophilic process can be dealt with using similar methods. These included carbon filtration, reverse osmosis or coagulation and flotation. The pre-rinse effluent from the post-removable hydrophilic process cannot be dealt with using these methods, and is best treated using coalescing filters or coagulation and flotation.

11.5.1 Carbon filtration

Carbon filtration is widely used to remove contaminants from solution in water. This complex process has been known to chemistry for many years. Laboratory use of carbon for purification was widespread in the last century, and industrial use of activated charcoal for purification of water has been established for 70 years in Europe and North America.

The carbon used comes from coal, peat and other sources. Carbon must be 'activated' before it can be used to remove contaminants from water. Various methods are used, all of which leave the material with a labyrinth of pores which gives the carbon an enormous surface area with respect to its bulk.

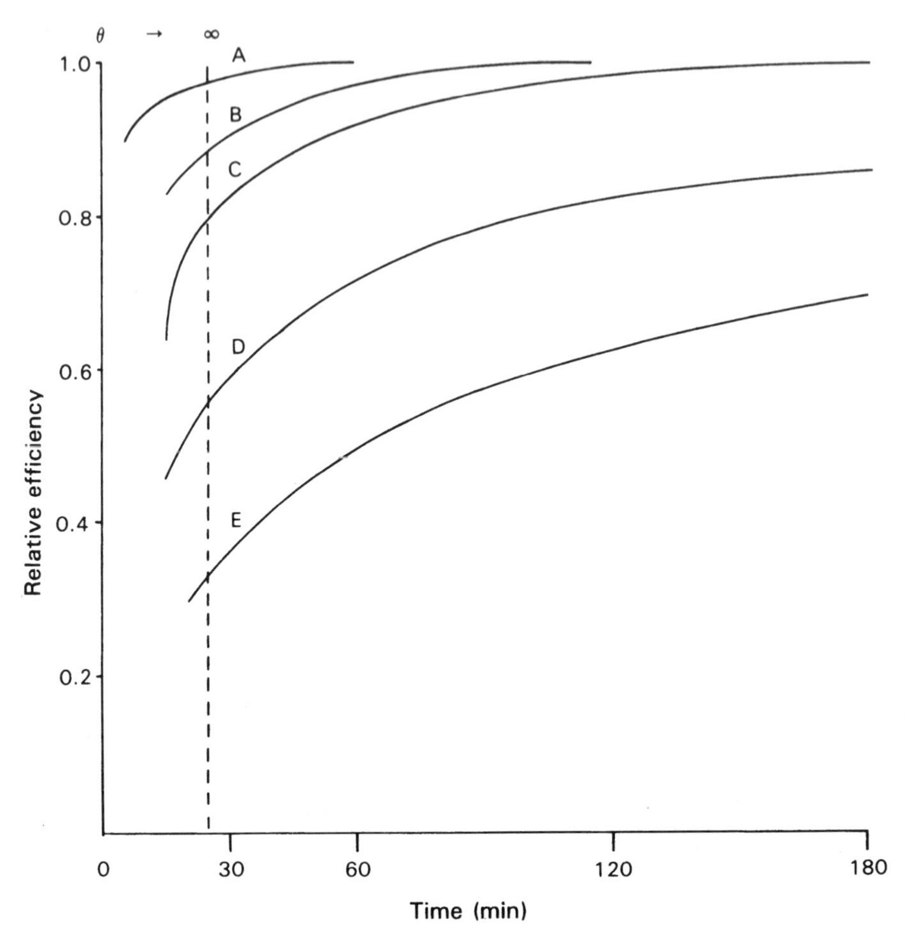

Fig. 11.1 Adsorption isotherm showing the influence of time and particle size on efficiency: curve A, 75–100 μm; curve B, 100–150 μm; curve C, 150–250 μm; curve D, 250–500 μm; curve E, 500–1000 μm.

The surface of the carbon adsorbs molecules of contaminant while allowing water molecules to pass through. The efficiency of the process depends on the preparation of the carbon and its particle size, and the concentration of the contaminants and their chemical nature. It is very difficult to predict which particular activated carbon will be efficient in removing penetrant contamination and there is no substitute for experiment. Two considerations must be taken into account: the time taken for the carbon to adsorb the contaminant onto its surface and the extent to which it removes it. This can be assessed in a series of experiments. Figure 11.1 shows the effect of carbon granule size on the effectiveness of a single type of carbon. This series of graphs also shows that a contact time of more than 20 min is needed for the activated carbon to be effective. Other tests indicate that a carbon well suited to the contaminant can achieve a volume efficiency of over 60% in an industrial application.

The design of a carbon filtration unit will depend on the maximum flow rate of the effluent. The flow of effluent from a penetration installation is intermittent. The effect of this can be reduced by using a holding tank. However, the filtration unit must be large enough to deal with the throughflow. If the design uses gravity flow with the effluent applied at the top of the carbon bed, some provision for periodic agitation of the carbon is needed otherwise canals will form in it and efficiency will be lost.

11.5.2 Reverse osmosis

When two solutions of a chemical of different strength are separated by a semi-permeable membrane the solvent (in this case water) will cross the membrane and reduce the concentration of the stronger solution to that of the weaker solution. This is called osmosis. A semi-permeable membrane allows molecules of solvent to pass while preventing molecules of dissolved materials from passing. The transfer of solvent across the membrane creates a pressure which is called the osmotic pressure. If, in a system where osmosis occurs, an external pressure equal to the osmotic pressure is applied to the more concentrated side of the membrane the process stops. If the external pressure is then increased, reverse osmosis occurs and solvent molecules pass from the more concentrated solution and it becomes stronger. Eventually most of the solvent can be removed in this way and a very concentrated effluent obtained.

This process has been applied successfully to suitable penetrating effluent rinsings. A high water quality is achieved; however, the initial cost of the plant is high and the process is relatively slow. It is essential that the compatability of the membranes with the concentrated penetrant is checked before installation. The membranes will be in contact with effectively 100% penetrant on one side. They are expensive, and unnecessary maintenance has further cost as well as being a nuisance. A number of semi-permeable membranes which show good resistance to concentrated penetrant waste are available.

11.5.3 Coagulation and flotation (flocculation)

In this process the effluent is mixed with a chemical or a blend of chemicals which adsorb the contaminant and form aggregations which are called flocs.

Since the effluent from penetrant processes is often an emulsion, this treatment is more accurately described as demulsification followed by coagulation and filtration. The choice of coagulant or flocculant is generally specific to the actual penetrant in use because demulsification must take place before flocs can be formed and filtered off physically. If coagulation is chosen as a method for dealing with penetrant process waste water, care must be taken in the choice of coagulant/flocculant if the water is to be recycled in any way. Some coagulant/flocculant chemicals contain chemicals which are not acceptable in any but the lowest concentrations (some as low as 50 ppm) for contact with components during penetrant processing. The best known of these are chlorine and sulphur, but some authorities object to others such as fluorine, sodium and potassium. This restricts the use of aluminium sulphates, alum, aluminium chlorohydrate, calcium chloride, chlorine, ferric chloride, ferric sulphate and sodium aluminate to installations where the treated water is sent to waste without recirculation. Care must also be exercised when choosing one of the polyelectrolyte types of coagulant which are now widely available to ensure that they do not contain any objectionable ions. Manufacturers of the polyelectrolyte coagulants are understandably reluctant to give too much detail about the exact structure of their products, but they should be able to confirm whether or not they contain significant quantities of sulphur, chlorine, fluorine, potassium or sodium. Some commonly used coagulant/flocculant chemicals do not contain the objectionable ions. These include calcium oxide and calcium hydroxide, but unfortunately these do not break the non-ionic emulsions of penetrant in water and consequently show little success. Coagulation and flotation (flocculation) works quite readily on physical mixtures of oil-based liquids in water where there is no emulsification and so such systems can be used to deal with the pre-rinse waters from the hydrophilic penetrant process. It is not possible to predict which coagulant will work in an individual instance and this must be decided by experiment.

Where a suitable flocculant system can be established it is possible for both the emulsified penetrant effluent and the pre-rinse waters to be treated in the same installation. The coagulant or coagulant mixture must be capable of demulsifying as well as forming flocs which can be filtered off physically.

11.5.4 Coalescing filters

Coalescing filters will not break emulsions of penetrant and water, and so are useless for treatment of waste from water-washable penetrant processes or lipophilic post-removable penetrant processes or the second wash from hydrophilic post-removable penetrant processes. However, they are very efficient from the treatment of pre-rinse water from the first wash in the hydrophilic post-removable penetrant processes.

Coalescing filters consist of a fine mesh or matrix of fibres. The ideal microfibre material is glass, but other chemically inert materials have been used successfully. The liquid droplets, which are suspended in another liquid (in this case the penetrant and water), become trapped on the fibres of the filter and run together to form ever larger droplets which are forced by the flow of liquid on to the downstream surface of the filter. The large drops of penetrant separate from the water and usually rise to the surface as most penetrants have a lower specific gravity than that of water. Where the coalesced liquid has a specific gravity greater than that of water it will sink on separation. It is clearly essential to keep the incoming effluent separate from the coalesced mixture, and care must be taken to ensure that the flow rate through the filter allows time for the liquid–liquid separation to be achieved. It is also essential for the effluent to pass through a pre-filter to trap any solid particles of dust present. Otherwise the coalescing filter will act as a physical filter and become blocked. Coalescing filters will have a mesh size which depends on the intended use. Mesh sizes of around 10 μm have been found to be satisfactory for dealing with the pre-rinse waters from the hydrophilic post-removable penetrant processes. Use of a 25 μm pre-filter normally ensures a satisfactory coalescer element life.

An advantage of using a coalescing filter for the treatment of pre-rinse water is that the recovered water can be recycled through that stage of the penetrant process many times as no chemicals have been added nor is there the possibility of accumulation of chemicals which may impair the performance of the process while the coalescing filter is operating correctly.

11.5.5 Biological contamination of effluent treatment plant

The mixtures of water and emulsified penetrant at very low concentrations which are typical of the processes offers nutrients for a wide range of micro-organisms. The low levels of oil-based penetrant in pre-rinse waters from the hydrophilic penetrant process may not be enough to deter some forms of bacteria and even fungi. Few of the air-borne and water-borne micro-organisms which have easy access to effluent treatment plant for penetrant processing waste would survive any significant time in the very high concentrations of penetrant which exist during reverse osmosis or in coalescing filters. Similarly, the chemicals used in coagulation and flotation (flocculation) do not offer a friendly habitat for bacteria and fungi, and there is also a fairly rapid throughflow. However, carbon filtration plants are a different matter.

In order to ensure a sufficiently long contact time with the effluent, carbon beds must be fairly large, typically 1.5 m high and about 1 m in diameter. Apart from the periodic agitation of the carbon, the bed can remain in place for 6 months or more. The temperature of the carbon bed rises as it is used, and as it becomes more impregnated with the organic molecules of the penetrant, detergent and emulsifying agent it becomes very attractive to bacteria and other micro-organisms which colonize it and multiply very quickly. The evidence of this is the foul smell of the water coming from the filter. This problem can be dealt with by using biocide which can be metered upstream of the filter or added regularly as a periodic dose. Where water is recycled it may be advisable to arrange for microbiological monitoring of the water from carbon filters. Many bacteria and other micro-organisms are acceptable in small concentrations but are potentially harmful in large concentrations. An imbalance of microbe population in water recycled from a carbon filter may even be a health hazard to operators. Certainly, where such water has a foul smell it should not be recycled through spray apparatus.

11.6 MANAGEMENT OF WASTE PENETRANT CHEMICALS

The problem of dealing with waste penetrant chemicals is quite distinct from that of dealing with effluent from the penetrant processes. The two separate problems arise together when the problem of dealing with the concentrated penetrant waste from reverse osmosis and from coalescing filters must be resolved.

It is good practice, and some specifications demand this, periodically to allow the penetrant to settle for 24 or even 48 hours and then to drain off the clean upper volume into clean containers and separately drain off the lower 10 cm (4 inch) layer to waste. On a busy installation this control may be needed every 6 months or even less, whereas on an installation which is used intermittently or relatively little the frequency can be reduced to an annual operation. The reason for doing this is to ensure that dust, rags, plastic cups, water and other contaminants are removed from the penetrant on a controlled and regular basis.

The operation obviously creates waste material. After application of penetrant to components a penetrant contact time is required, and during most of this time penetrant drains from the surface of the component. Unless drainage takes place over the penetrant reservoir it is not acceptable for the drainings to be returned to the reservoir. This also creates a volume of waste material. When the lipophilic post-removable penetrant process is used there are two drainage stages, i.e. after penetrant application and after application of the emulsifying agent, and such processes involve two sources of waste which are not water based. From time to time accidents occur. All fluorescent penetrants appear quite similar on superficial inspection as do all colour contrast penetrants. It is not difficult for a maintenance operator to top up a reservoir with the wrong penetrant. When this occurs, which thankfully is not often, no amount of readjustment will satisfy a quality auditor. It may be possible to make a mixture resulting from such an accident work to an extent but no quality auditor will accept it. The only proper solution to accidents is to reject the mixture as waste and start again with new supplies of material.

The situations outlined so far create waste which essentially contains no water and such waste is treated on that basis.

The use of detergent solutions (hydrophilic removers) has grown to such an extent over the past 25 years world-wide and over the past 35 years in Europe that the lipophilic post-removable process is in retreat world-wide and is almost unknown in Europe. However, it has created a volume of waste material which must be dealt with. The solution of detergent (hydrophilic remover) contains up to 95% water and becomes spent owing to the chemistry of the detergent removal process. The frequency with which such solutions are changed depends on how busy the installation is. Clearly an installation which is in constant use for 24 hours a day will need frequent changes, whereas one where the process is used only from time to time will need only occasional change. Both types of installation will cause effluent which is up to 95% water.

Another potential source of water-based effluent is the use of aqueous developers whether of the solution, suspension or colloidal type. As in the case of hydrophilic removers, the concentration of chemical in the water is low, – 60–150 g/l is typical – and the frequency of changing such developers depends on the frequency of use of the installation. A third type of waste product is spent dry powder developer. It is particularly important, if dry powder developers are to maintain their potential performance, that in a powder storm or other application which involves re-use a small quantity is used and replaced frequently. This creates a quantity of solid waste material.

11.6.1 Penetrant waste materials which can be burnt

The word 'incinerated' is often used in technical literature for burning. Although incineration is only another word for burning, it implies that such burning is done in a controlled way so as to minimize any environmental threat. The following waste penetrant materials can be disposed of by controlled burning: penetrant drainings, the bottom 10 cm from penetrant tanks (provided that it does not contain too much water) removed during the 6 monthly cleaning of penetrant tanks, drainings after application of lipophilic emulsifying agent, the concentrated contaminant from reverse osmosis, the non-water-based phase from coalescing filters, contaminated penetrant and spend lipophilic emulsifying agent. Many penetrants contain only carbon, hydrogen and oxygen in significant quantities, and so on total combustion they produce CO_2 and water. The dyes used to make penetrants visible, whether by direct inspection or inspection under UVA when the penetrants fluoresce, will contain a small amount of nitrogen so that on burning a small quantity of nitrogen oxides is produced. Some penetrants, mostly the non-water-washable types, contain very small amounts of phosphorus and so oxides of phosphorus will be produced when these materials are burnt. On burning, oxides of nitrogen and oxides of phosphorus must be scrubbed out. It would appear that with the current alarms about CO_2 levels in the atmosphere it may become necessary to scrub CO_2 from the gases produced leaving only water. From a chemical point of view it is not difficult to remove CO_2 from gases but it may offer some interesting challenges to chemical engineers.

11.6.2 Dealing with solutions of waste in water

Aqueous solutions of waste materials arise from spent detergent (hydrophilic) remover solutions and spent aqueous developer. Spent aqueous developer may be in the form of a solution or a suspension. The suspension can be filtered and the solid dealt with separately from the liquid. The solutions from developers may well contain traces of nitrite or chromate (whether from aqueous suspension or aqueous solution developers) which must be dealt with specifically. Aqueous solutions can be dealt with using carbon filtration.

11.6.3 Solid waste

Solid waste arises from spent developers. Spent dry powder developer waste is one source, and when aqueous suspension developer waste is filtered solid waste remains. Similarly, the non-aqueous suspension developers leave a solid waste after evaporation of volatile solvent. Fortunately, natural minerals and some carbohydrates have proved to be excellent raw materials for penetrant developers. Therefore disposal of solid waste developer material is not a serious problem. However, care must be taken with disposal of dry powder developer waste on account of the very small particle size of the material, which ranges typically from submicron sizes to 40 μm.

11.7 GENERAL SUMMARY

Increased concern over the environmental impact of industrial chemicals has been a feature of the past 30 years, and controls have been imposed and will become more thorough with time and experience. It is very difficult to think of any reason to regret this trend as the pressure on the ability of land, water and air to absorb effluent increases. Legislation and codes of practice for control of release of industrial effluent already exist in many countries. Two areas are already important and will become unavoidable. One is consultation with various authorities, e.g. water authorities and environmental production agencies, and the other is maintenance of records. It will become commonplace for the supplier of penetrant materials to have to liaise with customers and official environmental protection agencies such as water authorities to ensure proper control of effluent and disposal of waste materials. Naturally, confidentiality will need to be respected, and it should be possible for manufacturers of penetrant materials to disclose formulation details to official environmental agencies in confidence so that the requirements of effluent treatment for the user can be defined. These may well vary from one location to another as the pressure on the environment or its vulnerability to the effects of industrial wastes of all kinds must be considered.

Everyone prefers a clear-cut directive which is universally applicable. Such situations leave little to opinion. Unfortunately, the considerations of industrial effluent are complex and do not allow a simple blanket regulation. Many factors must be taken into account and the judgement of the local official environmental protection agency sought. An open approach by the user of the penetrant materials, the supplier of those materials and the local official environment protection department(s) will save many possible problems.

QUESTIONS

1. Some chemicals can be classified simply as 100% biodegradable: true/false.
2. Biodegradability is

   (a) a simple measure of a chemical which can be defined as straightforwardly as its boiling point
   (b) a simple measure of biological oxygen demand
   (c) estimated from measurements involving about ten different aspects of the process
   (d) a characteristic of a chemical which can be measured by one well-defined method

3. Chemical oxygen demand is

   (a) a measure of all organic material in a sample
   (b) a measure of all material present in the sample which can be oxidized using the chosen oxidizing agent
   (c) a method of estimating the available oxygen in a sample
   (d) a method of estimating the halide ion concentration in a sample

4. Effluent from the penetrant processes may be

   (a) emulsified rinsings in water
   (b) a mixture of penetrant and water in two phases
   (c) vapour from volatile organic solvents
   (d) all of these

5. Emulsified rinsings of penetrant in water can be treated by

   (a) physical filtration
   (b) activated carbon filtration
   (c) a coalescing filter
   (d) none of these

6. Coagulation and flotation are sometimes used to treat emulsified penetrant rinsings. Some coagulants fail to work because

   (a) they do not mix well with the effluent
   (b) they are not readily wetted by the effluent
   (c) they fail to break the emulsion
   (d) they do not ionize sufficiently in the effluent

7. Coalescing filters
   (a) are ideal for separating penetrant from water in effluent from the pre-rinse stage of the post-removable processes
   (b) can be used for rinsings from the water-washable penetrant processes
   (c) will treat effluent from the post-rinse stage of the lipophilic removal process
   (d) benefit from regular back flushing

Note: There is only *one* correct answer to questions 1–6. Tick your choice and check it with the correct answers on p. 216.

Answers to questions

CHAPTER 2

1 (d); 2 (c); 3 true; 4 (b); 5 (a).

6. A defect is some discontinuity which makes the component or material unfit for its intended use.
7. Mention should be made of the effect of repeated loading on a existing crack in the surface, or any vulnerable point, in causing the defect to grow relatively rapidly until the material is too weak to withstand the load. When a static load is imposed on the material the effect is normally much slower.
8. Any break in the integrity of the surface of a material is a stress raiser. It may be a defect, but it may also be a bolt hole, a change in section or a point where a stiffener is added.
9. (a) Initial freezing of the metal
   (b) Primary processing
   (c) Secondary processing
   (d) Service
10. Any four from
    (a) crater cracks
    (b) lack of fusion
    (c) undercutting
    (d) lack of penetration
    (e) cracks in the weld metal
    (f) cracks in the heat-affected zone

CHAPTER 3

1 (d); 2 (b); 3 (a); 4 (b); 5 (e).

6. The oil and whiting method used a penetrant (the oil) a remover (soapy water) and a developer (whiting).
7. Post-emulsifiable penetrant contains no emulsifying agent so that penetrant

trapped in discontinuities resists over-washing with water. Emulsifying agents are poor solvents for dyes and so greater intensity of colour or fluorescence could be achieved.

8. Water-soluble developers leave a uniform layer over the surfaces of components provides that areas of entrapment are drained so that the problem of heavy layers of developer at corners, threads and keyways is avoided.

CHAPTER 4

1 (c); 2 (b); 3 (d); 4 (d); 5 (a); 6 (c); 7 (c); 8 (d).

9. The emulsifying agents, which are known as lipophilic removers, mix with the surface excess penetrant to form a water-washable solution. The remover contact time must be controlled very closely to prevent the emulsifying agent from mixing with penetrant in defects before the final wash. The detergent or hydrophilic remover solutions do not mix with the penetrant on the surface but change the surface free energies of the surface material and the penetrant so that the two separate and the penetrant becomes stabilized in the detergent (hydrophilic) remover solution. Provided that the concentration of the remover solution is within the recommended range and the temperature of the solution is not higher than 40 °C (104 °F), the effect on penetrant entrapped in defects is very small.
10. Non-aqueous suspension developers contain a volatile organic solvent which is good solvent for penetrant. When this type of developer is applied the solvent mixes with the penetrant, ensuring contact with the developer particles, and draws the colour of the penetrant into the developer layer. The great danger is that the layer of developer particles applied is too thick and the indication becomes masked or is spread too thinly to be interpreted or in some cases even seen.
11. The volatile organic solvents which are used to remove surface excess penetrant dissolve the penetrant very easily and quickly. If they are applied at the removal stage of the process by spray, flooding or immersion, it is likely that penetrant will be washed out of defects. Even when one of these processes is shown to work for an application the removal must be automated to ensure that the same conditions are used each time. The wipe technique does not carry the dangers of the other methods of application and can be used manually.
12. Manufacturer's recommendations and specification concentrations for hydrophilic remover solutions must be followed for two major reasons. First, the remover and the penetrant are designed to work together as a pair and the concentration of the remover solution is an integral part of the process. Second, the activity of the remover does not increase directly with concentration but changes slowly until a point is reached where the activity increases very rapidly with small increases in

concentration. Maximum concentration set by manufacturers and specifications are well below the point of change.

13. Penetrant processes are classified as follows:

    (a) whether indications are shown under white light or UVA (black light), i.e. colour contrast or fluorescent;
    (b) according to the method used for removal of surface excess penetrant.

14. Liquid penetrant testing can be used to inspect any essentially non-porous solid surface regardless of shape or size. It can be applied to a wide variety of materials. However, it is essential to check compatability with plastic, resins and composite materials. The process will only provide indications of defects and other discontinuities which are open to the surface and free from contaminant. There is also a limit to the surface roughness which can be accommodated. When very rough surfaces are inspected only coarse and moderately coarse defects can be indicated. However, this limitation is true of all NDT techniques for finding surface defects.

CHAPTER 5

1 (a); 2 (d); 3 (c); 4 (c); 5 (d); 6 (c); 7 (a); 8 (b); 9 (a); 10 (c).

11. Mechanical methods of pre-cleaning surfaces before penetrant testing will cause the openings of surface-breaking defects to become peened over. This may close the opening to the defect completely and at best reduce it. The only exception to this is when a special procedure is used with a fruitstone grit. The effects of peening over can be overcome by an etch. However, this may not be acceptable as in many cases the material loss due to etching is not allowed and hydrogen embrittlement may occur in some metals after contact with etchant chemicals.
12. While it is true that any method of application of penetrant which ensures that the whole surface to be tested is covered is acceptable, great care must be taken to adjust the technique for removal of surface excess penetrant to allow for the thickness of the penetrant layer on the surfaces. The layer of penetrant left after application by immersion, flow-on or brushing of penetrant is much thicker than that left by conventional spray or a fog application, while electrostatic spray leaves a very light covering. The thicker layers of penetrant will need more vigorous removal than the lighter layers.
13. Water washing is used

    (a) for removal of surface excess water washable penetrant
    (b) for the pre-rinse stage when detergent (hydrophilic) remover solutions are used
    (c) for the post-rinse after application of detergent (hydrophilic) remover solutions

(d) for washing away the mixture of lipophilic emulsifying agent and penetrant at the end of emulsifier contact time in that process

In all water-washing procedures the following must be controlled: the distance between the spray nozzle and the surface of the component (30 cm minimum), the water or air–water pressure (2.4 bar maximum), the water temperature (ideally between 15 and 30 °C (60 and 85 ° F) and the length of time of the wash (less than 2 min).

14. The commonest method of applying lipophilic emulsifiers is for the components to be immersed in a bath of remover at the end of the penetrant contact time. This is followed by a period of drainage. Great care must be taken to rotate components during this time to ensure regular emulsification of surface excess penetrant, and the contact time is critical. Over-emulsification leads to loss of indications.
15. Ideally surfaces are dried in a stream of warm mobile air. The temperature of the air should be high enough to ensure that large and complex parts are dry within 10 min; 55–65 °C (130–150 °F) is enough for components up to 100 kg (210 lb) mass. Clean dry compressed air at 0.2 bar pressure can be used to remove water from threads, keyways and small holes before components are put in the drier. Immersing components in hot water at around 80 °C (176 °F) for 5 s before they are passed to the drying station is also a useful aid to drying.
16. Etching components may be forbidden if loss of material is unacceptable and hydrogen embrittlement is feared.
17. Electrostatic spray application of penetrant is very useful when large components of simple shape are to be inspected since this avoids the need for unduly large tanks of penetrant and the problems of immersing large components without splashing and other practical problems. This method of application is also useful when hollow parts with small holes are to be inspected. Turbine blades and stators often have cooling holes on or near the edges of the aerofoils. When electrostatic spray is used much less penetrant has the opportunity to enter these holes and bleed back later. The available penetrant is electrically charged, and a Faraday cage effect which actively prevents large quantities of penetrant from passing into the hollow area occurs in small apertures. This reduces bleed-back so much that the areas between cooling holes on turbine blades can be inspected.
18. Solvent removers are very aggressive, and if they are applied by spray, immersion or flow-on they will remove not only the surface excess penetrant but also penetrant trapped in shallow and fine defects, so thus seriously reducing the sensitivity of the process. Provided that flooding is avoided, the wipe technique with a cloth or tissue wetted with the solvent does not cause loss of performance.

19. The chance of seeing a fluorescent indication relies on the intrinsic fluorescent brightness of the material, which cannot be influenced by the viewing conditions, and the intensity of the UVA (black light) illumination and the contrast conditions, both of which can be influenced by the viewing conditions. Generally, a minimum intensity of UVA (black light) of 1000 μW/cm^2 is needed at the inspected surface –some specifications demand higher levels. The darkness is measured as the level of ambient white light acceptable which is less than 20 or even 10 lux (2 or 1 ft candle). It is also necessary to measure the white light from the UVA (black light) bulb. When higher levels of white light are present, this light competes with fluorescent indications making it progressively more difficult to see. They may even disappear.
20. Thixotropic penetrants are very useful when a small area on a structure or component is to be inspected, particularly if it is overhead or in an awkward place. The one special point about application of this type of penetrant is that the gel must be broken down by smearing out or stippling, otherwise sensitivity is lost.

CHAPTER 6

1 true; 2 (c); 3 (d); 4 (b); 5 (d); 6, order 2, 3, 7, 1, 4, 6, 5.

CHAPTER 7

1 (c); 2 (a); 3 (a); 4 (b); 5 (c).

CHAPTER 8

1 (b); 2 (d); 3 (c); 4 (b); 5 (a); 6 (b).

CHAPTER 9

1 (c); 2 (a); 3 (b); 4 (d); 5 (a); 6 (c); 7 (c); 8 (d); 9 (a); 10 (c).

CHAPTER 10

1 (d); 2 (a); 3 (b).

4 The high pressure mercury vapour arc bulbs used in inspection lamps emit energy throughout the ultraviolet, visible and infrared spectra. The Wood's glass filter cuts out wavelengths below 310 nm which include hazardous wavelengths in the UVB and UVC regions of the spectrum.

Cracked filters will allow these dangerous wavelengths to pass and can easily harm inspectors.
5 (c); 6 (d); 7 (a); 8 (c); 9 (a).

CHAPTER 11

1 false; 2 (c); 3 (b); 4 (d); 5 (b); 6 (c); 7 (a).

Index